Homeopathy Reconsidered

Natalie Grams

Homeopathy Reconsidered

What Really Helps Patients

Springer

Copernicus Books is a brand of Springer

Natalie Grams
Heidelberg, Germany

ISBN 978-3-030-00508-5   ISBN 978-3-030-00509-2   (eBook)
https://doi.org/10.1007/978-3-030-00509-2

Library of Congress Control Number: 2018954617

Translation from the German language edition: *Homöopathie neu gedacht: Was Patienten wirklich hilft* by Natalie Grams, © Springer-Verlag GmbH, Germany 2018. All Rights Reserved.
© Springer Nature Switzerland AG 2019

This Copernicus imprint is published by the registered company Springer Nature Switzerland AG
The registered company address is: Gewerbestrasse 11, 6330 Cham, Switzerland

Preface to the English Edition

This book is not only a story but it also has a story. It is not actually a scientific publication but is based on scientific thinking and principles. It is not a biography but it does also tell a personal story. I would like to explain that in this preface.

The book was first published in German in 2015 and was intended for all those interested in homeopathy, especially the practitioners, to express certain theses in which I reject the untenable parts of homeopathy but try to show a way to integrate the parts of it that seem to be worth preserving into modern medicine. With this book, the fundamental discussion about homeopathy as part of the health system and its claim to be medicine within medicine has been brought to life again in Germany—with unexpected intensity. And this has affected me personally—I suddenly stood in a prominent position, both with regard to media interest in the newly created public debate and in the sharp, sometimes very personal criticism I received from the homeopathic scene—which was the original target group for this book.

I was a free-practising homeopathic physician in Heidelberg from 2009 to 2015. During my medical studies, I was already not only convinced but honestly enthusiastic about homeopathy. I leapt on the first opportunity to take over a homeopathic practice. I was quite successful in my profession and my patients gave me a lot of positive feedback.

That is the beginning of my personal story, and it could have gone on like this, but…

One day I was interviewed by a journalist who was researching for a critical book on homeopathy. Actually, I was convinced that I had represented homeopathy well in the interview. Because of that, I was utterly speechless when I finally held the published book in my hands, for it spared no criticism of homeopathy—without appreciating anything of what had seemed to me in the interview to be an excellent defence of the method. Spontaneously, I decided to come up with an alternative draft and began to work my way into the literature and scientific study and assessment of homeopathy.

Well, to cut a long story short, it was incredibly difficult for me but as I went ahead with this work, more and more illusions about homeopathy which I had hitherto harboured began to fall away from me every day. I gradually came to realize that there were no sound or convincing reasons for it and its assumptions. I also had to realize that science, in its rejection of homeopathy, was not playing a conspiratorial role but took on the issues at hand with great seriousness and impartiality. That's how my book came about—but it went in a completely different direction to the one originally intended: a very critical direction. In this book, I tried to justify the worthlessness of the clearly untenable parts of the homeopathic theory and to leave them behind me. But I was so attached to the method at that time that I put a lot of effort into working out positive, medically relevant features, and gave suggestions for how one might make it usable for daily medical practice on a scientific basis.

For me personally, this meant abandoning my life plan. While I was still working on my book, I decided with a heavy heart to close my private homeopathic practice. In some passages of the book, I still speak in the tone of the practising homeopath—but at the time of printing, it was over. It was a very hard time, both on the way to this decision and afterwards, in many ways. Today I work in science communication, for the German Skeptics Society and as a freelance author.

From my current point of view, however, the book still seems to me almost like a defence of homeopathy, although it clearly rejects the notions of the principle of similarity, of increasing the effectiveness of remedies through potentiation into the literally infinite, or of the spiritual vital force in man and the spiritual medicinal power in the remedy. But my attempts to save the idea of intensive attention to the patient, through conversation which is not limited to the physical symptoms, constitute an essential part of my reflections in the book.

And I still feel obliged for that today. Even if homeopaths never use my theses for a discourse that might ensure the survival of homeopathy as a medical method. On the contrary, among homeopaths today I am considered a *persona non grata*, a traitor. They refuse to move away from the untenable view that homeopathy can provide a specific drug therapy.

But something good has also come out of all this because it has allowed me to move forward by constantly broadening my horizons as I read further studies and engage in many discussions with competent and well-meaning people from medical science, science in general, and the world beyond.

So this is the explanation for my introductory sentences. This book is a piece of personal history, and perhaps even a small piece of contemporary history. But it also expresses an admiration for science, reason, rationality, and not least honesty, all of which we owe to the patient in the medical profession, in addition to our expertise and sense of care. I am very pleased now to be able to present my book in English. I hope that it will give you the essential facts about homeopathy in a comprehensible way, but also and above all, through the example of my own history, what it means and how important it is to free yourself from deep misconceptions and prejudices, even if this may affect your own life plan.

The book should be readable for anyone interested, despite its scientific focus. Beyond the bibliography and references, I have therefore deliberately used references to and quotations from Wikipedia where possible. The Wikipedia references serve only as a quick introduction to terminology that some readers may not be familiar with. They are not meant as sources providing evidence in the strict scientific sense.

I have added another chapter to the original version of this book to tell you the "whole story", including what has changed and evolved since the German edition was published.

I wish you an inspiring and thought-provoking read - and good health!

Heidelberg, Germany   Dr. Natalie Grams
April 2018   http://www.netzwerk-homoeopathie.eu
                                          http://www.homöopedia.eu
                                             http://www.skeptiker.de
                                         http://www.natalie-grams.de

Acknowledgements

I am very happy that this book has been translated into English. I would like to thank everyone involved in its creation. In particular, I would like to thank Udo Endruscheit for making the English text better than the German text ever was. Thanks to Springer Verlag, Heidelberg, for continuing to stand behind the book's message, and especially to Angela Lahee, Frank Wigger, and Stephen Lyle. I thank my friends from the Homeopathy Information Network, GWUP (German Skeptics Society), Konsumentenbund, and the Giordano Bruno Foundation for all their support.

Contents

1

How and On What Basis Does Homeopathy Treat?

Under homeopathic therapy, I have seen severe anxiety and depression disappear, malignant cancerous ulcers recede, and acute purulent tonsillitis cured.

And yes, I am fully aware that homeopathic medicines (globules) do not contain anything that can be held responsible for this effect - unless you ignore all the common laws of science.

I'm a doctor, so I studied medicine. And I've been a convinced homeopath for a long time. However, as a doctor, I am a scientist. Because of this, I was no longer able to live well with homeopathy, whose principles somehow *feel* good, but whose principles *completely contradict scientific thinking*. I also thought this was hardly a responsible attitude toward my patients. I lacked conclusive explanations about the mechanism of action and proof of effectiveness of homeopathy. With this book I would like to explore this gap in more detail and encourage a new dialogue.

I have written it as a kind of memorandum to make it clear that some of the points are my own thoughts and conclusions that remain to be discussed. It is not therefore a purely scientific treatise, even though in this book science will have much more say than it has done so far in homeopathy. The path was difficult for me, because it meant dealing with my own professional, but also ideological principles. Maybe you'll feel the same when you read it.

The starting point is this: every day, patients came to my practice and reported with emotion and relief that their complaints had improved since the beginning of treatment. And it was not always just a slight cold. No, I would treat patients with problems of severe addiction, anxiety, and depression, some of whom hadn't been able to live a normal life for weeks. I would treat patients

© Springer Nature Switzerland AG 2019
N. Grams, *Homeopathy Reconsidered*,
https://doi.org/10.1007/978-3-030-00509-2_1

who had been undergoing therapy for years - be it psychological or conventional medical treatment: patients with cancer and other chronic diseases such as asthma, neurodermatitis, chronic inflammatory intestinal diseases, allergies, sleep disorders, pain, etc. How could they be helped by a method that has been proven to prescribe "nothing"? This question preoccupied me in my professional life, and I tried to pursue it as a physician trained in the natural sciences but also simply as an open-minded person. It was a big step for me, as an avowed homeopath, to realise that, despite the successes mentioned above and the great demand, there seem to be hardly any rational arguments in favour of homeopathy.

The fact is that proponents of the method believe, against all reasonable arguments, in the effects of the white beads full of nothing, and they see their ideas as being sufficiently confirmed by the success of their treatments. When asked how the effects are to be explained, they are either evasive or turn all the principles of logic and science upside-down in their argumentation. Perhaps the resistance of homeopaths to the demand for scientific proof of efficacy is so great because they think they can establish it by inspection, merely by observing that "homeopathy works". Critics and opponents of homeopathy, on the other hand, consider these recorded treatment successes to be an error of faith, something that is not based on a principle of cause and effect, and that can only be explained by the good old placebo effect. Some of those critics do not necessarily find fault with this, as long as it does not prevent the right medical measures from being taken if danger is imminent. At the end of the day, they just appeal to homeopaths to provide reliable evidence for this anecdotal efficacy and to explain the mode of action.

On the one hand, it is a fact that many patients turn to homeopathic treatment and claim that it helps them. On the other, I can also confirm that there is a whole range of cases where homeopathic treatment has had no effect at all. And that it even failed to achieve a placebo effect. One of my teachers of homeopathy who is well known for his treatment successes once said in a seminar: "If the effect of homeopathy were based solely on the placebo effect, then my success rate should be 100% - because the patients come to me from far away with great expectations, as a last resort, and under the pressure of great suffering; they have to wait a long time for an appointment, but I then devote all my time and ability exclusively and intensively to them, for a number of hours. However, I only reach 50% - so it must be something else."

Unfortunately, he did not say what might have been the reason for this, nor how he would explain the effect otherwise (apart from the error in reasoning that a placebo effect must occur in 100% of the treated cases). For me this remained unsatisfactory, and I had to ask myself this: if it's clear that there is no active ingredient and above all no "energy" in the homeopathic medicines to which one can attribute an effect, why do so many people still benefit from treatment based on such a (nonsensical) method? What is true about Hahnemann's theories that they are still so persistent, contrary to all reason? When did he develop homeopathy and how has medicine progressed since then? Which parts of his construct can still be justified in the 21st century? Why do patients continue to turn to homeopathy in such large numbers? Where is homeopathy vulnerable and where can we learn something from it? Where is homeopathy really nonsense? And where is our conventional medicine a nonsense of another kind?

According to Nobel laureate Daniel Kahneman, there are two ways of thinking: intuitive, automatic, fast thinking and conscious, rational, logical, laborious, slow thinking (Kahneman 2012). When they think about it intuitively and quickly, many people simply find homeopathy good. But what are the specific features that lead to this good feeling? And how can we, with our much slower scientific thinking, put such a feeling into figures, data, and facts so that both can be satisfied - the good feeling of so many patients who turn to homeopathy, but also the science that underlies modern medicine today?

To begin with, let me give an example of the course of a homeopathic treatment. Mrs. M. has been telling me about her persistent back pain for half an hour. I confine myself to listening, taking notes, watching the patient. When she stops, I encourage her with a simple "Tell me more". After some time listening and observing, I notice that certain topics come up more than once. On several occasions she reports a feeling of "being tied up" and having a stiffness in her back when the pain is very severe. She feels trapped or as though tightly gripped by something, and the worst situations are when she is resting, at night, or when she can't move. Subsequently, she reports that she felt trapped in her marriage, which had been a painful experience, and that it had taken her a long time to free herself from it. Somehow, she also connects the pain with her marriage. Yes, when she thinks about it now, her whole life has been influenced by a hitherto unconscious feeling of being *cramped and constrained*, of being *unable to move*. She has always been better off outside, where she is able to take in some fresh air. This tends to counter bad moods and back pain. Right now, at this very moment, she has the feeling that the pain has subsided, because she has realized that somehow it has always been about these topics in her life. She has been stressed by this all along. In her childhood she was rather wild and restless, always happy to move around. Since she has no longer been able to do this, she has felt restricted and discouraged - and completely ossified.

About an hour has passed at this point, and I have not made any findings or carried out any examination. An orthopaedic colleague would already have taken care of this, but unfortunately without being able to find a clear cause of the complaints. As a homeopath, I only try to find a kind of individual guiding principle, an individual basic theme, a red thread, or certain particularly unusual utterances in the expressed complaints, so that I can place the symptom in an individual context corresponding to the given patient.

Here I must emphasize that I am a "classical" homeopath. In classical homeopathy, which goes back to the German physician Dr. Samuel Hahnemann, the aim is to find *one* cause for the complete symptomatology of the patient, and to prescribe *one* piece of medication according to the criteria of the homeopathic method. It is not so much about general symptoms such as "back pain", but rather about conspicuous features that distinguish the entire symptom picture of one patient from that of the next patient.

In the case of Ms. M., I notice that she changes from the symptom of back pain ("stiff, tied up, worse at the beginning of a movement, then better, worse in rest, improved in warmth") to a topic that seems to be universally valid for her: "rigid, stiff, gripped, cramped, painful", contrast with "free, moving, pain-free". This consultation involves a homeopathic technique of anamnesis which requires training and which I will elaborate on in later chapters.

I prescribe a homeopathic medicine from the family of Anacardiaceae (*Rhus toxicodendron* or poison ivy), which is suitable according to the thinking underlying this method (Sankaran 2003a, 2003b, 2005). I give it to Mrs. M. directly in a high potency (C200) as globules and give her more globules for use at home.

Four weeks later I meet Mrs. M. again and ask her about the course of events. She reports that right after the last conversation she felt much better, that the pain was acceptable; a week ago the pain had become more severe again, so she then took the globules I had given her. Thereupon, her situation improved again. She told me that she was really satisfied with the treatment, as she felt more relaxed, more alive, and stronger. The pain hardly played a role in her life any more, and she had begun to defend herself more actively against the circumstances of her life which she perceived as a limitation. She feels this to be the real benefit of the therapy, because it has reduced her overall stress and she can turn to new things in her life.

What has happened? As a homeopath, I would say that I have successfully prescribed a homeopathic remedy that solves Mrs. M.'s problem. But is that true? How could a plant of the Anacardiaceae family be related to this change, and how is there supposed to have been any healing? From a medical or phytotherapeutic point of view, poison ivy is not known as an analgesic. The biological textbooks say that, at best, these plants can cause a sensation of stiffness, tightness, and pain after skin contact. They cause a local sensation. But how could this have been conveyed to my patient, especially as nothing from the original plant was contained in the prescribed potency?

The simplified, common explanation of homeopathy for this claims that everything is connected in a "spirit-like" way with everything else, and this is how the necessary healing information or sensation of poison ivy is "transferred" to Mrs. M. via "energy". The healing information has been "shaken out" of the poison ivy during drug production by "potentiation" and is now available for healing according to the principle "similar cures similar" (in this case, "stiff sensation when touching the plant" and "stiff feeling in the case of back pain"). Despite all the happiness I may feel about the result, I am aware that my method is in many respects contrary to valid physical, chemical, biological, and medical principles. A simple acceptance of the idea of a transferable "energy" would violate the principles of reason and science. And it would remain unclear how a cure could have been achieved in this way.

So what did happen then? First of all, Mrs. M. got time to open herself up to me with all her worries and private feelings. I took the time needed to let her finish talking until we came up with a principle that seemed to be coherent and meaningful to her: "I always feel so stiff and locked up and yet I want to be free and move." This connection seemed to explain something to her about herself that she was not yet aware of. Apparently, she felt an immediate relief. It seems that my patience, my candour, and - not least - my specific questions, posed according to the homeopathic method at certain points, had touched her emotionally.

Mrs. M. had thus succeeded in finding out something about herself. Somewhere far removed from the pure symptom of back pain, she gained an awareness of herself that affected her imagination. She experienced this self-awareness as an immediate general relaxation, including in the relevant physical area. The patient attributed the physical improvement automatically to the globules. This seemed to be repeated even after taking it again. She could do something herself and was no longer at the mercy of her pain. Moreover, the insight she had had in conversation with me seemed to enable her to initiate a change in her life that was appropriate to the situation, which meant that she felt that her overall stress level was diminished.

In contrast to the normal medical approach, which would presumably have been the prescription of some pain medication, homeopathy strives for a more far-reaching approach:

- Time for the patient
- Openness and understanding (even for unusual and "peculiar" things)
- The possibility of expressing physical, emotional, and spiritual things and putting them into context (holistic)
- An individual approach (not just any pain, but *my* pain)
- Improve self-perception and self-awareness and initiate situation-oriented changes in life

- Medication with a high placebo effect - probably depending on the form and the number of medicines to be taken (Buckalew and Coffield 1982).

Of course, one wonders whether homeopathy is really needed for all these factors. Wouldn't a sensitive and calm conversation with a normal doctor have the same effect? With some common sense? Sure! Unfortunately, modern medical structures provide hardly any physicians with such a possibility. And many men in particular find it difficult to go to a "psycho doc" with "only bodily diseases", to talk about larger, more individual contexts. Hence, in the best case, homeopathy provides what many patients long for: individual, empathic counselling with the aim of seeing each patient as a human being in his or her own right, without putting a "psycho-stamp" on the person or administering unnecessary medication just to show that they have done something.

Perhaps this was the great achievement of Hahnemann, who pioneered homeopathy. In fact, it was by no means customary in his day to practice such medicine. Bear in mind that the first patients Hahnemann thought were confirming his theory had long conversations with him every day for weeks to months. And he omitted the then common treatment methods such as phlebotomy (bloodletting). Who knows whether the conversation and avoidance of this pointless weakening were not actually fully sufficient to make the patient healthy? Any "*energy*" in the given globules would then be unnecessary - then as now. Thus, one could say that homeopathy is a method that involves having calm conversations that enable the patient and his or her body to do something for themselves. I see the globules as placebos, but also as carriers of an individual autosuggestion, which I shall explain in detail later.

But how exactly does homeopathy actually achieve this? Are there differences from normal conversation and psychotherapy - and from a normal placebo effect? Why and under what circumstances could homeopathic medicine be effective, even if its drugs don't contain any active ingredients or a specific "energy"? And how do such considerations match with science-based medicine, which is not founded on supernatural considerations?

References

Buckalew L, Coffield KE (1982) An investigation of drug expectancy as a function of capsule color and size and preparation form. J Clin Psychopharmacol 1982:245–248

Kahneman D (2012) Thinking fast and slow. Penguin, reprint edition

Sankaran R (2003a) An insight into plants vol. 1 (Anacardiaceae). Homeopathic Medical Publishers, Mumbai

Sankaran R (2003b) The spirit of homeopathy, 3rd edn.Homeopathic Medical Publishers, Mumbai (1999)

Sankaran R (2005) Sensation refined. Homeopathic Medical Publishers, Mumbai

2

Homeopathy - What Are We Talking About?

2.1 Is There "One" Homeopathy?

This question is easy to answer: No, "homeopathy" is not a term with a uniform meaning.

If someone mentions the term "homeopathy", it is assumed that everyone knows what is spoken of and what is meant by it. A big mistake. Homeopathy is by no means a closed system or a uniform method, and it is also often combined or confused with other methods of so-called alternative medicine. That is why I would like to clarify at the outset what we are talking about when I refer to homeopathy.

Founded by Dr. Samuel Hahnemann around 1796, this special healing method evolved in different directions from the very beginning. Today, there are many different schools, methods, and partial methods as well as many fare-dodger variants and fashion trends in Germany and worldwide under the broad, comprehensive name of homeopathy. The so-called "classical" homeopathy goes back directly to Hahnemann, but there are also different variants here; "genuine" homeopathy refers strictly to Hahnemann's texts. Other forms are, for example, complex-agent homeopathy, quantum-logical homeopathy, and psychological homeopathy. In addition, there are some proposed healing methods related to homeopathy that use at least similarly manufactured medicines (e.g., anthroposophic medicine, Schuessler salts, Bach flower remedies). In addition, homeopathy is often mixed with other alternative healing methods such as Tibetan massage, electroacupuncture, colour therapy, and aromatherapy or singing bowl therapy, to name just a few.

© Springer Nature Switzerland AG 2019
N. Grams, *Homeopathy Reconsidered*,
https://doi.org/10.1007/978-3-030-00509-2_2

I would like to emphasize once again that "*homeopathy*" as such does not exist as a single well-defined method. This makes it so difficult to assess homeopathy and get a clear picture of it. There are many different "homeopathies" with few principles in common. In addition, just as there is no such thing as "*homeopathy*", the terms "*homeopath*" or "*proponent of homeopathy*" are equally ill-defined. From the absolute hardliner to the slight doubter to the sceptic, many variants are possible - and this is another reason for the differences in the way homeopathy is seen and practiced. This in turn also contributes to the fact that it is difficult or even impossible to make a comprehensive assessment of homeopathy.

Viewed from the outside, homeopathy seems to be more coherent and uniform than it actually is. Within homeopathy there are great differences and sometimes even contradictory views, whether it is about conducting and interpreting a homeopathic anamnesis, the selection and dosage of medication, or additional recommendations for the patient during homeopathic therapy. Basically, this makes it possible for anyone who deals in homeopathy to act as a homeopath and to give homeopathy their own "individual spin". And of course, every homeopathic school claims to be the best, if not the only valid one.

The label "homeopathy" is not associated with a quality criterion. In Germany, for physicians who wish to work as homeopaths, there has been a protected additional designation since 2003, preceded by a standardised training course, and there are now also training regulations for non-medical practitioners. However, the training is also freely accessible to laypersons, and what they have learnt can be practised at their own discretion. Homeopathy as a healing method is thus offered and applied in many different forms and with widely varying quality.

> Homeopathy is a healing method, not a professional title. Doctors, non-medical practitioners, but also lay people can practise it. A non-medical practitioner is not necessarily a homeopath (unless he or she specializes in homeopathy). This is often passed over in everyday language.

To clarify what I am talking about when I use the term homeopathy in this book, I will give you a brief introduction to the origins, principles and application of homeopathy today. I adhere strictly to Hahnemann, as his statements form the basis for all kinds of homeopathy and we will thus at least remain close to a general, fundamental assessment.

2.2 Samuel Hahnemann, Founder of Homeopathy

There is a lot of literature about Dr. Samuel Hahnemann, the German doctor and pharmacist who founded homeopathy. At this point, I will not go into the details of his life, since there are plenty of good accounts of this subject to which I would have nothing to add. There is also a wealth of autobiographical material from his own pen. I deliberately quote Hahnemann himself in many places. It takes some practice to get a feeling for his old-fashioned way of writing, but it is instructive to get to know him better and become acquainted with his rather special manner.[1]

I will confine myself to a few key facts. Samuel Hahnemann was born in 1755 as the son of a porcelain painter from Meissen, Saxony. He studied medicine in Leipzig (Saxony) and Erlangen (Bavaria) from 1775. In 1810 he published the first version of his fundamental work on homeopathy, "*Organon der Heilkunst*" - "*Organon of Medicine*" (from here on I will refer to this simply as the "*Organon*"). However, he also pursued several other professions and interests, changed his place of residence extremely frequently, married twice, and had eleven children - a life that perhaps looks a little unstable. Hahnemann died in Paris in 1843. On his tombstone, at his own request, the inscription reads: "I have not lived unnecessarily". In my opinion, two aspects of his biography are essential for the development of homeopathy:

- Obviously, Hahnemann was a restless spirit. His life was marked by many changes of address, phases of rich creativity, periods of deep poverty, various activities carried out with varying degrees of success in different professions and fields of thought: from writer to pharmaceutical manufacturer, from assistant of the personal physician of the Empress of Austria to translator, from Freemason to what may have been the first psychotherapist - and more.
- Moreover, he seems to have been a man who generally doubted everything that was taken for granted. With verve, he railed against traditional medicine and throughout his life, he repeatedly came into conflict with laws, superiors, colleagues, and just about any prevailing doctrine.

Taking into account everything I have read from his own pen and about him, Hahnemann seems to have been a wise mind, and an intelligent and very concrete original thinker who bravely and energetically opposed the

---

[1]For the literal quotations from Hahnemann's Organon, the translation by *Boericke* is used in this book (see the bibliography).

medical establishment of his day and unrelentingly condemned many causes for grievance. For example, he was one of the first doctors to advocate hygiene and his interest in the human psyche was unusual in those days. But he also seems to have been a highly idiosyncratic and self-righteous person who was not afraid of polemic when he expressed his opinions. In this book you will find quotes from Hahnemann again and again, and these will make it clear how complex his thinking was and how clearly he wrote everything down. The mode of expression is difficult and often only comprehensible after several readings, but I am always amazed at how concretely he was able to say certain things that are still valid today. However, other parts need to be discarded in the light of present knowledge; I will deal with this later. Above all, his striking character seems to have enabled Hahnemann to live up to his motto *Sapere aude*, borrowed from Kant ("*Answering the Question: What Is Enlightenment*", 1784*)*, and to go his own way in medicine. *Sapere aude* means "Dare to know" (Wikipedia, keyword: Sapere aude).

To this day it is unclear whether he also does justice to the second meaning or translation of this motto: "Dare to be rational/to use your reason". Hahnemann was then and can still now be described as an "awkward customer" and nutcase on the one hand, but a genius on the other. Is he rather to be understood as ingenious, or was his idea of homeopathy just humbug? In this book, I would like to try to make a clearer distinction between what is reasonable and justifiable in homeopathy from today's perspective and what needs revision or what is in fact outdated and must therefore be discarded.

2.3 Medicine in the Days When Homeopathy Came into Being

To understand how Hahnemann came to invent homeopathy and the associated way of thinking, it is important to recall the era in which he lived and the state of medicine at that time.

In those days, medicine was still characterized by mystical ideas and somewhat radical therapeutic methods. Indeed, in Hahnemann's time, medicine was primarily based on the bodily humors theory (humorism, humoral pathology). In short, this is the idea that four bodily fluids (blood, phlegm, and black and yellow bile) can be produced depending on which disease a person has. This concept was developed by Hippocrates (460–370 BCE) and remained the predominant doctrine in medicine until into the 19th century (Wikipedia, keyword humorism). Even though Galen (approx. 129–216 A. D.) thoroughly revised the humorism of antiquity, he maintained its principles, and the aim of any therapy was still to remove the "corrupted juices" from the body. This was done using drastic measures. In addition to bloodletting, the focus was on vomiting and laxatives. The excess of poor juices present in the event of an illness had to be excreted as blood, sweat, pus, or faeces. Drugs were administered as patches, enemas, oils, ointments, compresses, incense, odorants, potions, tinctures, pills, or powders. The measures were not very specific and, as I said, rather drastic, so that, expressed just as drastically, patients had the choice of surviving either the disease or the therapy. Many did not survive either. At that time, illnesses were much more likely to be life-threatening and often spread over entire regions in broad, epidemic waves (Wikipedia, keywords: history of medicine, humorism).

Hahnemann was not satisfied with the conventional doctrine that it was crucial to flush out or suppress. Above all, he was bothered by the fact that therapies were so unspecific and that they generally led to a weakening of the patient, which in many cases went beyond the already considerable weakening caused by the disease itself. In addition, doctors were considered much more as superior authorities than they are today, and it was not on the agenda to lend a sympathetic ear to the distressed patient. Hahnemann strongly resisted this view of the doctor-patient relationship and the prevailing doctrine (e.g., in regard of hygiene, which had not yet been practised at that time). True to his character, he raged against many of the prevailing doctrines of medicine in his fundamental work, the Organon.

Hahnemann's Organon consists of individual paragraphs, laid out like a patchwork, partly sorted according to themes, partly mixed up in a somewhat confusing way. In it, Hahnemann develops the principles of his own medicine, which he calls *Homeopathy*. The name "homeopathy" comes from the Greek *homousious* for 'similar' and *páthos* for 'disease' and basically means "cure similar with similar". With this method he consciously sought to distance himself from the traditional healing principle. He described the latter principle, which was based on the therapeutic doctrine *contraria contrariis curentur* (healing opposites with opposites), as 'allopathic medicine' with a defamatory intention.[2]

---

[2]As a German synonym for the term "allopathy", homeopaths introduced the term "Schulmedizin" in Hahnemann's day. This is linguistically the equivalent of "orthodox medicine" in English but has rather different connotations. The term "Schulmedizin" was - and still is today - used in a derogatory sense in relation to scientific medicine, in the sense of a "pure academic", "memorized", "rigid", "dogmatic" medicine. This may seem strange, because it is precisely methods such as homeopathy that are "dogmatic" and "rigid" in this sense, and not scientific medicine, which is constantly developing in a self-critical way. For this reason, I do not use the term "orthodox medicine" in my German-language publications when writing critically about homeopathy, but rather "normal" or "scientific" medicine.

All in all, medicine at the time was still much less a natural science than it is today - simply because scientific principles were only researched and clarified much later on. Magical thinking, amulets, talismans, healing waters at the fair, holy pictures, and religious and spiritual associations were at that time just as much part of medicine as the rather non-specific medical interventions mentioned above. Disinfection, hygiene, and physiological or biochemical processes in the human body were largely unknown. Remember that Hahnemann lived at a time when Virchow's cellular pathology was still a thing of the future; Virchow developed it around 1850. Hahnemann therefore knew neither about the cardiovascular system nor about body cells and did not know that functional disorders at this level could be a cause of illness. Virchow's discovery was ground-breaking and finally led to rejection of the Bad Juice Theory (Humorism) - which Hahnemann was still living with as a doctor - after 2500 years. It was only around 1860 that the physician Semmelweis discovered the principles of bacterial infections and finally the basis of microbiology, which was greatly expanded by Koch's and Pasteur's insight that diseases could be triggered by viruses and bacteria (around 1876). The first antibiotic was only discovered in 1897 and it took until 1928 before Fleming used it medically. Many other important milestones in modern science and medicine were also only reached later (Wikipedia, keyword History of Medicine). Many similarly momentous developments occurred in mathematics, physics, chemistry, and other fields in the 19th century, along with scientific methodologies and methods of detection such as statistics or the principle of causality and evidence-based research.

Our conception of humankind has also undergone enormous changes since Hahnemann's times. While many phenomena could only be explained with vague ideas and myths in those days, there are many insights today which allow other, more apposite conclusions (Schmidt-Salomon 2014). Particularly since the development of modern naturalism in the 20th century, there have been sufficient scientific explanations available to completely dispense with the likes of miracles, myths, and otherwise inexplicable aspects of humans, their abilities, and their illnesses.

Today things look like this:

In philosophy, naturalism is the "idea or belief that only natural (as opposed to supernatural or spiritual) laws and forces operate in the world." Adherents of naturalism (i.e., naturalists) assert that natural laws are the rules that govern the structure and behaviour of the natural universe, that the changing universe at every stage is a product of these laws. (Wikipedia keyword: Naturalism, section philosophy)

Hahnemann developed his theories at a time that is considered pre-scientific in today's sense. And so, in many respects, he may just have been a child of his time and could not have known better. Today, however, there are different principles in science and medicine which I will elaborate on in more detail.

2.3.1 The Homeopathic Method - What Is Different?

Then as now, Hahnemann's homeopathic approach differs markedly from scientific medicine:

The principles of homeopathy according to Hahnemann

- The patient is seen primarily as an individual. Homeopathy is not about the treatment of symptoms per se.
- Instead of looking for an external cause of a disease, the doctor's attention is focused as precisely as possible on recording an individual medical picture and an internal disposition. (There are no bad juices or external influences that make us ill. This view is still valid today.)
- The body has a self-healing potential that can be stimulated. (There is nothing to flush out or fight against.)
- The doctor focuses on getting a complete picture of the illness and the patient's frame of mind, and the drug-finding process derived from it is highly individual and "distinctive" - i.e., it differs from patient to patient. (The doctor must therefore listen attentively to the patient.)
- The symptoms are only a reflection of the actual inner problem: the disturbance of the "vital force". The overall portrayal of the symptoms results in a complete "patient picture", which is to be compared with a homeopathic "drug picture".
- It is just a detuning of this "vital force" that leads to illness. It is pointless looking for external causes. Rather, the sole task of the physician is to find a remedy *similar* to the diseased condition. This means that such a *similar* drug should cause an analogous kind of "artificial disease" in the body. In this way, the body recognizes the (similar) real illness and is thus able to heal itself.

> "Every effective drug excites a kind of its own disease in the human body, the more peculiar, excellent and violent the disease, the more effective the drug is. In the disease to be cured (...) one uses the medicine which is capable of causing another, preferably similar, artificial disease and which will be cured; Similia similibus. (Similar things will be healed by similar things.)"[3]
>
> "By observation, reflection and experience, I discovered that, contrary to the old allopathic method, the true, the proper, the best mode of treatment is contained in the maxim: *To cure mildly, rapidly, certainly, and permanently, choose, in every case of disease, a medicine which can itself produce an affection similar to that sought to be cured!*" Hahnemann (1997), Organon, Introduction to the 6th edition

Hahnemann imagined things in that way: Belladonna, or deadly nightshade, if inadvertently taken by a healthy person, causes confused, high-fever conditions with sweating, red face, great fatigue, and often fear of death. However, if taken in very small doses for therapeutic purposes, Belladonna should be able to cure a sick condition with similar symptoms. In this way, the sick person's condition is "held up in front of a mirror", and this almost empowers the body to initiate the steps required to heal itself.

Hahnemann was convinced that he had discovered this principle in a self-experiment in which he took the herbal medicine Cinchona (cinchona bark), which was already in use at the time. The self-experiment is difficult to replicate, because Hahnemann himself only incompletely documented it. But the following is said to have happened: malaria (formerly called "alternating fever"), a disease already known in Germany at the time, is associated with quite typical symptoms (periodic fever, severe diarrhoea, and significant weakening of the organism). Hahnemann believed that his self-experiment had shown that taking Cinchona caused malaria-like symptoms in himself. This led him to the fundamental idea of homeopathy. He asked: "If you give a malaria patient with typical symptoms a cure that is able to cause similar symptoms in a healthy person, wouldn't this have to be compensated?" And indeed, what he had thought up so theoretically seemed to work in practice.

---

[3]Hahnemann S.: Experiment with a new principle to find the healing powers of the medicinal substances, together with some insights into the previous ones. ("*Versuch über ein neues Princip zur Auffindung der Heilkräfte der Arzneisubstanzen, nebst einigen Blicken auf die bisherigen*") First publication in Hufeland's Journal of Practical Medicines, 2nd volume, 3rd part, 1796. Hahnemann (1755–1843) formulated at this point for the first time the principle of homeopathy, which he later founded with his main work "Organon der rationellen Heilkunde" (1810), namely the principle of similarity: *Similia similibus curentur* (similar things may be healed by similar things).

There followed many years of experimentation and refinement of his homeopathic principle.

Then as now, critics doubt Hahnemann's interpretation of his self-experimentation. The experiment could never be replicated (Hopff 1991). Hahnemann may simply have suffered an allergic reaction to the quinine contained in the cinchona bark (Wikipedia, keyword Homeopathy; Aust 2013). I will deal with the principle of homeopathic drug testing in a later section.

During his lifetime, Hahnemann examined 27 other remedies by taking them himself, precisely recording all the symptoms that emerged as a result - and from then on prescribing them for similar clinical pictures. Hahnemann himself describes the basic homeopathic principle of similarity in this way:

> Now, however, in all careful trials, pure experience, the sole and infallible oracle of the healing art, teaches us that actually that medicine which, in its action on the healthy human body, has demonstrated its power of producing the greatest number of symptoms *similar* to those observable in the case of disease under treatment, does also, in doses of suitable potency and attenuation, rapidly, radically and permanently remove the totality of the symptoms of this morbid state, that is to say, the whole disease present, and change it into health; and that all medicines cure, without exception, those diseases whose symptoms most nearly resemble their own, and leave none of them uncured.
> Hahnemann (1997), Organon, § 25.

> The curative power of medicines, therefore, depends on their symptoms, similar to the disease but superior to it in strength (§ 12–26), so that each individual case of disease is most surely, radically, rapidly and permanently annihilated and removed only by a medicine capable of producing (in the human system) in the most similar and complete manner the totality of its symptoms, which at the same time are stronger than the disease.
> Hahnemann (1997), Organon, § 27

An exact patient picture is recorded in the homeopathic anamnesis. This is compared with the similar drug picture that has arisen from homeopathic drug testing. In the patient's body, the administration of the drug is intended to lead to a kind of artificial disease that enables the body to overcome the "right" disease itself. The artificial disease gives it the "tools", so to speak, to deal with the very similar real disease; I will go into this later.

It should be added that at Hahnemann's time quite a few physicians followed the principle of signature theory handed down by Paracelsus (1493–1541).

In short, this means that similar things in the universe correlate. This was also used for healing purposes; for example, beans should help against kidney diseases, walnuts against brain diseases. Hahnemann was not alone in upholding this idea at the time. Today, the theory of signatures is considered refuted and unusable for scientific findings (Wikipedia, keyword Doctrine of Signatures).

Even today, homeopathy differs from scientific medicine in its view of what should be considered and treated in the event of illness. In contrast to normal medicine, homeopathy covers not only physical, but also emotional and mental ailments. Above all, the inclusion of spiritual aspects is an unusual feature in comparison to the approach of normal medicine.

Another big difference between scientific medicine and homeopathy is the dosage of their medicines. Whereas a physiological effective limit is the measure of dosage in normal medicine, homeopathy claims to be successful precisely with non-physiologically effective dosages.

2.3.2 Homeopathic Repertories and Materia Medica

As a result of homeopathic pathogenetic trials (HPT) (previously referred to as homeopathic drug testing), first carried out by Hahnemann himself and later by his students, long lists of symptoms were drawn up which were expected to be eliminated with the assigned drugs, since they caused similar symptoms in healthy people. These lists are called *repertories*. Every homeopath still works with repertories today and there are many variants of them. In this respect, too, there is no uniform doctrine in homeopathy.

Let me exemplify with an extract from a randomly accessed online repertory for the symptom "pain, joints, rheumatic":

Aconitum Napellus, Antimonium Tartaricum, Arnica Montana, Arsenicum Sulphuratum Flavum, Aurum Metallicum, Belladonna, Benzoicum Acidum, Bryonia Alba, Cactus Grandiflorus, Calcarea Carbonica, Calcarea Phosphorica, Calcarea Sulphurica, Causticum, Chamomilla, Chimaphila Maculata, Chininum Sulphuricum, Cimicifuga, Racemosa, Cocculus Indicus, Colchicum Autumnale, Colocynthis, Dulcamara, Ferrum Metallicum, Ferrum Phosphoricum, Formica Rufa, Guajacum Officinale, Hepar Sulphur, Iodium, Kalium Bichromicum, Kalium Iodatum, Kalium Muriaticum, Kalium Sulphuricum, Kalmia Latifolia, Lac Caninum, Lachesis Muta, Lacticum Acidum, Ledum Palustre, Lycopodium Clavatum, Mercurius Solubilis, Natrium Muriaticum, Natrium Sulphuricum, Nux Vomica, Rhododendron,

Chrysanthum, Rhus Toxicodendron, Salicylicum Acidum, Sanguinaria, Canadensis, Spigelia Anthelmia, Staphysagria, Sulphur, Veratrum Viride.

Furthermore, descriptions of drugs, the so-called *Materiae medicae*, were created. They bring together a detailed collection of typical or important symptoms and characteristics that are established for the image of a given drug, along with texts containing interpretations and suggestions for use. These *Materiae medicae* indicate the patient picture, as recorded in the anamnesis, for which a homeopathic medicine should be helpful. Today, every homeopath uses these works.

Again, let me exemplify by an excerpt from a *Materia medica* that describes *Rhus toxicodendron*, the medicine prescribed in the discussion of Mrs. M:

This plant represents various symptoms. First, there are poisoning symptoms and skin problems. On the other hand, *Rhus toxicodendron* has a pronounced effect on ligaments and joints and acts on neuralgia. And finally, the clinical picture of restlessness, a strong urge to move, and insomnia is addressed. *Rhus toxicodendron* is particularly indicated when the corresponding symptoms deteriorate in cold and wet weather and in autumn and winter. Another sign that this remedy is indicated is that the symptoms are more pronounced at night than during the day.

Improvement of symptoms:

Improvements in symptoms are observed in warmth, hot baths, sweating, rubbing of the affected areas, and continuous movement, including stretching of the body and lying on the back. Continuous movement, changes of position at rest, warmth and warm, dry weather improve symptoms. Wrapping, rubbing and kneading, stretching of the limbs, and holding the affected body parts have a positive influence on the condition.

Deterioration of symptoms:

Deterioration is observed when lying on the side and at rest. Increased discomfort is caused by draughts, soaking and dislocation, overstretching, and hypothermia. Deterioration is caused by cold and damp in any form, and especially at night. The symptoms also get worse in autumn and winter. Cold after sweating, cold washing, cold food, cold drinks and overexertion also have a worsening effect.

Indicated physical symptoms for the use of *Rhus toxicodendron*:

• General stiffness especially in the morning

- Rheumatism and rheumatoid arthritis
- Tendon and muscle strain
- Arthritis
- Sciatica
- Shoulder-arm syndrome
- Torticollis
- Lumbago
- Neuritis and neuralgia, especially sciatica
- Bullous dermatitis and pustular eczema
- Polydermia
- Skin rashes with blistering
- Herpes
- Common cold
- Fever and chills when hands are exposed

Since Hahnemann's death, both works have been continuously supplemented and extended by various homeopathic investigators and authors, so they are by no means uniform. On the contrary, *Materiae medicae* and repertories with different focal points are created and used by different homeopathic schools. True wars of faith arise around the "correctness" of one or the other variant. Today *Materia medicae* and repertories are also available as computer databases with search functions that allow one to search for drug images or symptoms; there are also different variants here.

From the outside, homeopathic medicine may be perceived as more closed and unified than it really is. I therefore stress once again that there is no such thing as a single, well-defined homeopathy, and nor can one speak of "the" *Materia medica* or "the" repertory (Wikipedia, keyword: Homeopathic Materia Medica).

2.3.3 Homeopathic Anamnesis

To find a drug similar to the disease condition, the first task is to record and describe that condition as precisely as possible. The homeopathic anamnesis is intended to serve this purpose. There are essentially four ways to obtain a homeopathic anamnesis:

Types of homeopathic anamnesis

- Broad anamnesis (e.g., classical homeopathy, homeopathy according to Kent):

The homeopath tries to record as many symptoms, particularities, and modalities as possible (what makes it better, what makes it worse?). This also includes the personal history and the family anamnesis. The therapist will ask for symptoms other than the main symptom, as well as general dislikes and preferences. The aim is to create an overall picture that is as detailed as possible.

- Deep anamnesis (e.g., homeopathy according to Sankaran):

Based on the main symptom, the homeopath tries to detect a deeper individual pattern in the patient and to bring this into connection with a generalizable impaired basic sensation.

- Anamnesis of peculiarities (e.g., homeopathy according to Boenninghausen, Sehgal, Scholten):

The aim is to identify some highly individual peculiarity of the patient. It may be an unusual or specific modality (what makes it better or worse), an absolutely typical constellation of symptoms (king-pin), or any kind of clearly prominent mental beliefs. ("King-pin" is a special term of the Sehgal method, which describes just these "individual peculiarities" as a single central "marker". Sehgal explained the "king-pin" also as the "main disposition" of the patient or "identifying marker" for finding the right homeopathic therapy.)

- Anamnesis by *Quickfinder*:

Although this procedure is probably the most frequently used, especially by laypersons, it has little to do with homeopathy in Hahnemann's sense. Symptoms or groups of symptoms are looked up using *Quickfinders*, or attempts are made to assign certain types of constitution to "suitable" homeopathic drug images (e.g., the sulphur type). In my opinion, this form of anamnesis must also include questionnaires, however detailed they may be. However, a questionnaire cannot replace a comprehensive homeopathic anamnesis. At best, it can only supplement it.

The therapist must invest a considerable amount of time for the first three methods. Maybe this is also the reason why the short-cut via *Quickfinder* is often used. The aim of all forms of anamnesis is to relate the symptom to the individual patient. For back pain, for example, the homeopathic anamnesis is based on questions such as these:

- What is the difference between this back pain and the back pain that I have treated this week in other patients?
- How does this patient describe his/her pain in contrast to others?

- Which image does this individual sensation and description resemble?
- Which patient picture comes up here, and which drug picture is similar?

In our initial example with Mrs. M. the following picture emerges. The chronic pain improves with warmth, worsens at the beginning of a movement, and then slowly improves; in the morning it is worse. She feels stiff and cramped by the pain. The following description would reveal a completely different picture: "Acute, throbbing, intense red area of pain; extreme pain with anxiety and restlessness; wants to lie covered until she sweats; but can hardly move in pain, otherwise she will feel nauseous and dizzy (belladonna)". Or otherwise: "Long-lasting dull pain, with the feeling that she cannot move, otherwise she will break; great inertia and tiredness (remedy: thuja)."

Of course, we also cover various aspects of back pain in normal medicine. However, we do this only to discriminate between an acute or a chronic picture and then determine the appropriate therapy (painkillers, physiotherapy, surgery, etc.). The picture in normal medicine is far less complex, and the questioning is far less detailed, simply because it is usually limited to symptoms.

The aim of homeopathic medicine is to reconcile a comprehensive, typical, and profound picture of the patient with a similar picture of a drug. The homeopathic anamnesis serves this purpose.

The anamnesis technique is by no means uniform, differing fundamentally across different homeopathic schools. While it is partly a more or less systematic general inquiry into current and past symptoms, findings, sensitivities, or family characteristics, some homeopaths may try to identify specific peculiar characteristics of the given patient. This may be a purely physical symptom on the one hand, or a mental problem on the other. Hahnemann only gave instructions to question the patient in detail and let them talk until they have said everything, but he did not specify any particular technique for questioning or anamnesis. Consequently, it is not possible to check in homeopathy whether an anamnesis has been made incorrectly or correctly - simply because there is no criterion. Of course, every homeopathic method claims to do this more correctly than the others.

2.3.4 Homeopathic Medicines (Potentization)

Once the patient's homeopathic anamnesis (of whichever kind) has led to a precise assessment of their condition, the therapist must find and prescribe a drug similar to that condition (with a different procedure depending on the method adopted). The detailed overall picture, the deep individual pattern of sensations, or some distinctive specific feature will become the basis for the diagnosis, which will in turn lead to the drug.

In Hahnemann's day, the following were used as medicines:

- inorganic/mineral substances (e.g., mercury, arsenic, sulphur, calcium carbonate, various salts),
- plant-based substances (e.g., rhubarb root, arnica, calendula), and
- substances of animal origin (e.g., castoreum, musk, ambergris, but also animal excrements).

Thus, it was only logical that Hahnemann also made such substances the basis of his drug therapy, particularly as there were no chemically or synthetically produced drugs yet. From this pool he drew his homeopathic remedies.

However, the problem was that Hahnemann occasionally felt compelled to administer poisonous medicinal plants to patients. An example is belladonna (deadly nightshade), which can lead to confusion and fever when ingested, but also to severe poisoning symptoms and death. He was confronted with the problem that he himself denounced: that he could only weaken a patient with such a poisonous remedy, even if the principle of similarity was fulfilled. This may have led Hahnemann to the idea of gradually diluting his toxic drugs until no more poisoning effects need be feared. Contrary to his expectation that the effect would now also have to weaken, he thought he detected evidence that it was actually improving. The healing results were even better when the drug was not only diluted, but also shaken, i.e., "dynamized and potentiated" (Organon, Paragraph 269). Hahnemann precisely specified this so-called potentiation (Organon, paragraph 270 ff.). I present it in simplified form as follows:

The potentiation of healing substances

An original substance (the homeopathic stock) in its pure form is dissolved and mixed in a ratio of 1:10 (decimal dilution, hence X- or D-potencies) or 1:100 (centesimal dilution, hence C-potencies) with a solvent (usually water or alcohol). According to Hahnemann, the diluted substance must now be treated by succussion so that it does not remain merely diluted. Dilution and succussion cause the "dynamization". "Succussion" means tapping the vial rhythmically on a "hard but elastic body" (Hahnemann, § 270 Organon). (Hahnemann himself recommended a "leather-bound book".)

In this way, a "potentiated" mixture - i.e., potencies 1X or 1C - is

---

> obtained, from which one takes a drop for a new succussion with a further nine (resp. 99) drops of solvent. This will result in the potency 2X (resp. 2C). By repeating the first process six times, one gets the potency 6X (resp. 6C).
> This drug solution obtained in this way is then usually applied to sucrose or lactose sugar globules. After the fluid has evaporated, it is ready for use in the given potency (as homeopathic globules).
> (Wikipedia, keyword Homeopathic dilutions)

Hahnemann reports that patients would not only be cured of their symptoms with such potentised drugs, but that they seemed to be strengthened internally, and came out of the illness with personal gain.

He tried to explain how such an immaterial drug could still have any effect at all and could lead to such surprisingly good results, given that a potentiated drug no longer contains any of the supposedly active material. For him, this was not necessary at all, because an illness is basically a "spiritual" process. It therefore needs something "spirit-like" to influence it. He describes it as follows in his Organon:

> Homeopathy can easily convince every reflecting person that the diseases of man are not caused by any substance, any acridity, that is to say, any disease-matter, but that they are solely spirit-like (dynamic) derangements of the spirit-like power (the "vital force") that animates the human body. (...) Hence Homeopathy employs for the cure ONLY those medicines whose power for altering and deranging (dynamically) the health it knows accurately, and from these it selects one whose pathogenetic power (its medicinal disease) is capable of removing the natural disease in question by similarity (simila similibus), and this it administers to the patient in simple form, but in rare and minute doses so small that, without occasioning pain or weakening, they just suffice to remove the natural malady whence this result: that without weakening, injuring or torturing him in the very least, the natural disease is extinguished, and the patient, even whilst he is getting better, gains in strength and thus is cured.
> (Hahnemann 1997, Organon, preface of the 6th edition)

The "spirit-like" aspect of the drugs was to be achieved through potentization. This was Hahnemann's idea:

> This remarkable change in the qualities of natural bodies develops the latent, hitherto unperceived, as if slumbering hidden, dynamic (...) powers which influence the life principle, change the well-being of animal life. This is effected by mechanical action upon their smallest particles by means of rubbing and shaking and through the addition of an indifferent substance, dry or fluid are separated from each other. This process is called dynamizing, potentizing (development of medicinal power) and the products are dynamizations or potencies in different degrees. Hahnemann (1997), Organon, Paragraph 269

> By means of this manipulation of crude drugs are produced preparations which only in this way reach the full capacity to forcibly influence the suffering parts of the sick organism. In this way, by means of similar artificial morbid affection, the influence of the natural disease on the life principle present within is neutralized. By means of this mechanical procedure, provided it is carried out regularly according to the above teaching, a change is effected in the given drug, which in its crude state shows itself only as material, at times as unmedicinal material but by means of such higher and higher dynamization, it is changed and subtilized at last into spirit-like medicinal power, which, indeed, in itself does not fall within our senses but for which the medicinally prepared globule, dry, but more so when dissolved in water, becomes the carrier, and in this condition, manifests the healing power of this invisible force in the sick body. Hahnemann (1997), Organon, Paragraph 270

By dynamization and potentiation a substance X should become spirit-like and can thus act on the spirit-like vital force. So much for Hahnemann's theory. He postulates this special transformation even for those substances that in themselves have no medicinal effect, for example table salt. For Hahnemann, this also solved two other problems:

Practical benefits of potentiation

- Toxic remedies could now be used with confidence.
- A patient was not weakened by such treatment.

That would certainly also have been desirable from an allopathic point of view. (I will discuss the problem of homeopathic potentization from a scientific point of view in detail in the next chapter.) But perhaps it was just these positive aspects, where he so clearly positioned himself with respect to the deficiencies of the medicine of his day, that made Hahnemann blind to the shortcomings of his theory. Even today, this is still a problem for medical progress: an apparently ingenious idea leads to a kind of blind spot for the shortcomings of medical progress. This is one of the reasons why medical and scientific research is so important today: it focuses on the search for blind spots of new and old theories (Kahneman 2012).

2.4   Homeopathic Diagnosis, the Principle of Similarity, and Homeopathic Drug Testing (Homeopathic Pathogenetic Trials - HPT)

Homeopathy is often accused of not making a diagnosis. This is not entirely true. In fact, it does not make a diagnosis that can be expressed in a single term, of the kind we are accustomed to in scientific medicine. The diagnosis in homeopathy is the patient's picture. This is made according to the different anamnesis principles which I outlined above. All the patient's statements about his or her symptoms, as well as the particularities of his or her life, character, mood, and feelings in the context of past or family complaints, etc., are integrated into an overall picture. Each patient always presents only *one* picture.[4] This one overall picture is then matched to a drug picture, which should in turn represent all these aspects. The drug image should be as similar as possible to the patient image. This is the basic principle of homeopathy: the principle of similarity.

---

[4]Cutting short this process of drug finding isn't actually allowed, but for simplicity's sake (especially for laypersons) it is often what is done in practice. Hence, medication A is given for symptom A and medication B for symptom B. However, Hahnemann's homeopathy was not conceived in this way, because homeopathy does not operate in terms of symptoms, but in patient and drug pictures. These are sometimes very complex and go far beyond a single symptom. For a patient (and his or her patient picture) there can only ever be *one* medicine in the sense of classical homeopathy.

> **The principle of similarity**
> Similarity in homeopathy means that a medicine that produces certain symptoms in a healthy person can cure a disease with similar symptoms or a similar condition. According to Hahnemann, this is done by causing a similar "artificial disease" in the patient, which is processed by the body. The latter can thereby overcome the actual disease.

Hahnemann himself describes the principle in his Organon as follows:

[…] that actually that medicine which, in its action on the healthy human body, has demonstrated its power of producing the greatest number of symptoms similar to those observable in the case of disease under treatment, does also, in doses of suitable potency and attenuation, rapidly, radically and permanently remove the totality of the symptoms of this morbid state, that is to say, the whole disease present, and change it into health; and that all medicines cure, without exception, those diseases whose symptoms most nearly resemble their own, and leave none of them uncured. Hahnemann (1997), Organon, § 25

A weaker dynamic affection is permanently extinguished in the living organism by a stronger one, if the latter (whilst differing in kind) is very similar to the former in its manifestations. Hahnemann (1997), Organon, § 26

The greater strength of the artificial diseases producible by medicines is, however, not the sole cause of their power to cure natural disease. In order that they may effect a cure, it is before all things requisite that they should be capable of producing in the human body an artificial disease as similar as possible to the disease to be cured, in order, by means of this similarity, conjoined with its somewhat greater strength, to substitute themselves for the natural morbid affection, and thereby deprive the latter of all influence upon the vital force. Hahnemann (1997), Organon, § 34

This means that a medicine that produces certain symptoms in a healthy person should be able to cure a disease.

The principle of similarity can be briefly explained as follows. Imagine your condition after drinking too much coffee. Symptoms such as palpitations, sweating, but also inner restlessness, tension, and nervousness appear. If you came to my practice as a patient, I might be able to recognize the medicine picture "coffee" (Coffea cruda) in your patient picture and administer it to you as a homeopathic medicine. In your body, the "coffee (overdose) disease" would then be simulated (artificial disease). According to Hahnemann, the body then initiates the necessary countermeasures to regulate the body and make it healthy again (self-healing). Thus, I wouldn't give you a remedy that slows down the heart rate (e.g., a conventional medical beta-blocker). Hahnemann states that otherwise the symptoms would only be suppressed, while the actual cause would not be treated.

Drug pictures were (and are) created as follows:

How drug pictures are created

A healthy person takes the drug (resp. the active ingredient), usually in an already potentized form, but occasionally also in the original form. Changes are now documented in the physical and emotional realm, but also in the sensory realm. This is called homeopathic drug testing, something Hahnemann explained in detail in his book "Reine Arzneimittellehre" ("Materia medica pura[5]"). I will come back to this later.

An assessment is made, depending on the clarity and frequency of the symptoms identified in the trial. Symptoms that turn out to be most typical and occur for almost all investigators receive a high level of valence, others a lower value. The results of the examinations are included in the repertories.

It should be said at this point that, despite all the best intentions of practicing homeopaths and the attempt at standardization, this is a very shaky procedure. There is by no means a uniform picture - much less than outsiders might think. There are dream trials, trials with homeopathic stocks, trials with already potentiated drugs, trials with purely female test groups, and more. Each trial group may come to a slightly different drug picture. It is true that only the characteristics with the greatest correspondence are then taken over, but there are nevertheless significant differences depending on the repertory, *Materia medica*, or homeopathic school! As a result, a homeopathic diagnosis is not always as clear as we would like it to be, or as homeopaths would like to present it. This is because opinions about test results, and hence also drug images, vary widely within homeopathy. As if this were not already

---

[5]Materia medica pura (Reine Arzneimittellehre), Translation by Dudgeon and Hughes, online: https://archive.org/stream/b28121612_0002/b28121612_0002_djvu.txt.

problematic enough, there is still a lot of scope for homeopaths to interpret their own individual cases of treatment.

A diagnosis is thus made in homeopathy. However, it is for the main part far less reliable and meaningful than in scientific medicine. This is just as big a problem in homeopathy as the theory of drug testing, which I will discuss later.

2.5   The Sensation Method in Homeopathy

Finally, to remove any ambiguity when I refer to homeopathy, I would like to clarify something. In my practice, besides classical homeopathy according to Hahnemann, I used a certain form of homeopathy: the so-called sensation method.[6] I used this method because it retains the basic features of the classical method, but can be used more structurally and in a more targeted way, and at the same time clears up many of the old myths. Other schools in homeopathy do not offer such a clear anamnesis and decision structure and allow the therapist too much purely subjective latitude.

From the very beginning, one thing that had always annoyed me in homeopathy was that there was no comprehensible basis for deciding on a drug. While therapist A had convincing arguments for drug A, therapist B was able to find a completely different drug for the same patient. What's more, a homeopath practicing for thirty years could know and prescribe a thousand medicines, a younger colleague perhaps only twenty. When, repertorizing (i.e., finding the remedy "similar" to the patient's picture) it was possible for something completely different to come out, even when the same therapist was at work, but the symptoms were weighted differently. So what medication was to be given? Which "diagnosis" was to be made after I had spent more than two hours with the patient?

Attending to a seminar, I received the answer to my question from a leading modern homeopath: there is a need to have an "instinctive feeling" and to trust one's intuition. After that, I started a desperate search for a method with a clear concept of drug determination. Since Hahnemann had stated that there is only *one* drug to cure each case (since a patient only presents *one* individual overall picture), then it must be possible to find it in a clear and comprehensible way. If there is only *one most similar* (*similimum*) among a wide range of similar drugs (*simile*), then it should be possible to

---

[6]Sensation method - that sounds rather unscientific. It's not a scientific method either! I will go into the benefits of this method later.

separate this clearly and logically from the others. With the sensation method developed by Rajan Sankaran, this wish seemed at first to be answered.

Dr. Rajan Sankaran is an Indian homeopath. He founded the sensation method because in many diseases the classical (homeopathic) anamnesis often reaches its limits. In particular, in the case of mental problems such as anxiety, neurosis, etc., it is often difficult for the patient to name specific symptoms, so the clinical picture remains unclear. With Sankaran's method the patient describes his or her suffering in images and metaphors on the emotional level, including descriptions of feelings, dreams, and the perception of his or her environment. In this way, an image of the patient is gained that makes clear a pattern in his or her life. This pattern might affect or might interfere with the patient or make him or her ill. The treatment then tries to address just this pattern. In this way the patient as a whole might be treated better than using a method solely based upon symptoms.

> **The basic idea of the sensation method**
> Many people experience their illness, but also their life situations, with a certain fundamental sensation. Usually they are not aware of this sensation. The sensation affects their experience of physical symptoms as well as emotional and mental challenges or stressful situations. Such a sensation runs through all dimensions of each individual human being. If this sensation becomes evident in the anamnesis, it helps the homeopath to find the most specific *similimum*, the most similar remedy. The sensation is the icing on the cake, so to speak, which condenses the patient picture into a "core sensation".

This core sensation can usually be expressed in one sentence. In our initial example: "I always experience myself so constricted and trapped and want to be free." This makes diagnosis much more precise. Thus, in a *Materia medica* of the sensation method, additional information can be found, apart from the medicine *Rhus Toxicodendron* mentioned in the introductory example.

The core perception could be this: caught, stuck, and held in a situation that needs to be escaped quickly. The central feeling revolves around being imprisoned, or stiffness, or somehow being restricted. It reveals a feeling of being held and the urge to move (Sankaran 2003).

Clear concepts for anamnesis management, case processing, and drug discovery made it possible for me to reach a less ambiguous diagnosis than with classical homeopathy. The sometimes very complex patient picture can be summed up in a core sensation. This made the method very valuable to me in the past and it was the basis of my homeopathic work. However, the most important thing for me nowadays is that this method offers a targeted approach to the mental complaints of a patient and allows him or her to identify a core problem. I'll deal with this in detail later. I found another advantage of the method in the instructions and courses on anamnesis management, which follow a clear concept. The anamnesis is perceived as something like tracing the "red thread" associated with the patient. This is done, for example, by describing body sensations, a very open questioning technique, reflecting the words used by the patient, and working with spontaneous hand gestures and doodles (spontaneous scribbling).

In this method, what the patient says is paraphrased (e.g., in the homeopath's own words: "You said you felt trapped?") or supplemented by open questions or encouragements ("Tell us a little more about the fact that you feel trapped?"). The aim of the anamnesis is to lead the patient to previously unconscious feelings. Because of this, the therapist tries to reflect the patient's statements with as little alteration as possible, and she or he should deliberately avoid intervention by steering. This is intended to ensure the greatest possible objectivity on the part of the homeopath, who refrains from expressing his or her own opinions and evaluations, while concentrating on and actively listening to the patient. A similar situation is known from psychotherapy, for example, in the method of active listening developed by Carl Rogers (Wikipedia, keywords: Active listening; Carl Rogers).

References

Aust N (2013) In Sachen Homöopathie – eine Beweisaufnahme. (The case of homeopathy - the evidence. 1–2-Buch, Ebersdorf (in German)

Hahnemann S (1997) Organon of Medicine, 6th edn (Translation Boericke). http://www.homeopathyhome.com/reference/organon/organon.html). Accessed 20 April 2018

Hopff W (1991) Homöopathie kritisch betrachtet. (Homeopathy viewed critically) Thieme, Stuttgart (in German)

Kahneman D (2012) Thinking, fast and slow. Penguin, reprint edition

Sankaran R (2003) An insight into plants, vol. 1 (Anacardiaceae). Homeopathic Medical Publishers, Mumbai

Schmidt-Salomon M (2014) Hoffnung Mensch. Eine bessere Welt ist möglich. (Hope on man. A better world is possible) 2nd edn. Piper, München (in German)

3

Is Homeopathy Part of Today's Medicine?

3.1 Why Do We Need Science at All?

Homeopathy is nowadays a more or less accepted part of medicine. More and more health insurance companies accept homeopathy as being reimbursable. But can we accept that? Is homeopathy really part of science-based medicine today? Part of an approach to health whose detection methods have been developed continuously since Hahnemann and are appropriate to the current state of research? I'd like to deal with this problem in this chapter.

Whether reading this book as an e-book or in a printed version, you are experiencing just now what we need science for. It has enabled us to develop and make things usable for everyday life: the electricity that is needed for the e-book, as well as the computer technology with which it was developed. The letterpress printing and the necessary printing machines are a result of scientific thinking. It follows, therefore, that science is by no means something abstract and unworldly, but a method that actual creates knowledge. Practical benefits can often be derived from this.

Science puts forward hypotheses. These are put to the test according to increasingly well developed and refined principles, which do justice to the specificity of their subject matter under investigation. The aim is to formulate hypotheses as simply and comprehensibly as possible and subject them to independent scrutiny - do the results speak in favour of the hypothesis or do they refute it? It often turns out that it is not necessarily the hypothesis itself that leads to further development, but rather its critical examination and the attempt to refute it. Sometimes hypotheses move outside common doctrine (like Galileo's world view, which was quite new at the time); but then all

© Springer Nature Switzerland AG 2019
N. Grams, *Homeopathy Reconsidered*,
https://doi.org/10.1007/978-3-030-00509-2_3

arguments in favour of this position should be clearly explained and independently examined (exceptional claims require extraordinary evidence). The point is not about being right, but about figuring out the truth or at least approaching it in a consensus. Science is so important because it can be used in many cases to decide whether a hypothesis, opinion, or idea is true or false, or express its degree of certainty (Wikipedia, keyword: Science).

It is not a matter of rigidly cementing knowledge once it has been gained. It is precisely the sceptical approach to supposedly already secured knowledge that drives science to question this knowledge time and again and to come up with the best solutions or definitions *at that point in time*, i.e., we conceive of scientific knowledge as being in constant evolution.

This scientific approach has given rise to the biological, chemical, physical, neurological, immunological, and other fundamentals of medicine, which today shape our understanding of disease and have led to well established and safe therapies. They are constantly evolving. ECG monitoring, X-ray, CT scanning, magnetic resonance tomography (MRT), antibiotics, microinvasive surgical techniques, sterilization and disinfection - without science and research we would not have all this.

Of course, there are also wrong conclusions and fallacies in this scientific research. If you consider how many biases and errors of thought are possible in this area, you may wonder how reliable results can be achieved at all (Herrmann 2013; Dobelli 2014; Wikipedia, keyword: Science). The way of science is complicated and agonizingly slow - but this procedure is the best way not to stay trapped by biased opinions and other mistakes. Science's mission is not to draw premature, intuitive conclusions, but to follow the slow, logical path of consistent reasoning (or refutation) (Kahneman 2012). Errors and misconceptions can - and will - occur at first, but later they can be reviewed according to new insights and corrected if necessary. It is precisely this principle of constantly questioning what has been tentatively accepted and adapting it to new knowledge that constitutes the very essence of science. Unfortunately, all too often we see that the knowledge acquired by science is also used in a pernicious way in practice. But the scientific approach is not to blame for that.

So what about homeopathy and science? The doctrines that Hahnemann set up in his Organon may well have been convincing and attractive in his day. But, of course, Hahnemann was limited by the state of science in his day. The methodologies that we consider scientific today did not exist at the time, neither in mathematics nor in physics, nor in the theory of medicine. Hahnemann published all his findings and conclusions. In his writings - the *Organon* and *Chronic Diseases* - he cites a whole series of reasons that led him to his conclusions. The fact that he sometimes reacted quite vehemently to criticism and didn't cultivate a particularly factual style of discussion is a different matter. He was so convinced that he was right that he simply thought any attempt at verification to be superfluous. This may have been acceptable in his time, but today we homeopaths still follow him in this. The effect of our globules is not so easy to explain, we claim, or we explain the effect in such a way that no one understands it. We do not take note of the fact that other causes or coincidence may be responsible for a change after taking the globules. We oppose an objective discussion of our ideas and principles. At the same time, however, we want our treatment practices to be part of medicine and we ourselves want to be recognized by it (and health insurance).

> The equation homeopathy = medicine = science is no longer valid today because the methods of science and medical knowledge have developed since homeopathy came into being.

If we want to do (more) justice to the equation, we must do the following: we must question our hypotheses and examine (or let others examine) what is right and what is wrong about them. We must put them up for discussion. There are points that have already been refuted; I will mention them. There are points which we may be able to reassess, and I shall also refer to them. There are points that could be positive. And there are points that I may not have thought of in my considerations. In any case, the path is long. I would like to begin this journey with my personal story in order to explain my motivation.

3.2 On My Personal Situation: In Conflict with Science

This section describes my own personal experience. I studied medicine because I was enthusiastic about the idea of accompanying people on their path to recovery. During my studies I was still inspired by this ideal. With increasing clinical activity, I became disillusioned by the way patients (people) were dealt with in everyday medical life. Even I myself had psychosomatic complaints after a serious accident, which could not be clarified or treated with conventional medical methods. I first turned to traditional Chinese medicine, then to homeopathy, which I already knew from a student group.

During my clinical work I was only able to practise homeopathy as a secondary subject. But at least I spent the weekends training in the method and its many subtypes and treating my own patients privately. A few years later I started my own business with a purely homeopathic practice. Here I was able to work the way I had always imagined: accompanying patients in their personal development. My dream seemed to have been realised. Unfortunately, however, my total conviction was soon to fade.

Since my studies, I'd always made every effort to improve my skills and knowledge within the method, and at first I had dismissed all doubts that came from outside: the whole world - and in particular homeopathy - could not be explained with purely scientific thinking after all! I adopted many theories of homeopathy, largely because I wanted to believe them. Firstly, this was my dream, and secondly, every day, I saw that patients benefited from the treatment. My doubts became stronger when the "scientific part" of my personality began to pay attention to background issues. I read the studies which were quoted so often, and a lot of other material and noticed that natural scientists didn't think as badly about homeopathy as homeopaths did about natural science. Unfortunately, scientists simply had better arguments. Please read the sources below yourself. I cannot explain it better than is already done therein.

It was a tough year. It forced me to acknowledge facts which were unfortunately not so bright. Actually, there are no scientific arguments *in favour of* homeopathy. And unfortunately again, one thing above all seemed to be true: there is nothing in homeopathic medicine that can be made responsible for any specific effect. Hahnemann's theory that there is an "energy" or a "spirit-like power" in such medicine is certainly not tenable. So why did patients still come to me? And, even more amazingly, why did they go away healthy?

My attempts to deal with this question have led to this book. What I have found in the literature so far has a familiar look about it: there is long-standing struggle between "It works! We don't know why, but we just know it!" on the one hand and "There's no substance, so there's no effect!" on the other. Convinced patients and therapists assume that homeopathy will be effective and see this confirmed by successful treatments. Scientists and homeopathy critics argue that drugs in which no active ingredient is detectable - and certainly no energy or spiritual power - can't possibly have any effect; the relationship between globule administration and a change in the patient could not be causal. We may only concede that homeopathy as an overall package has a positive effect by courtesy of a good doctor-patient relationship and the intense discussion that involves, while a general placebo effect could be accepted for the drugs (Ernst 2002; Shang et al. 2005).

Homeopathy has been in this situation since the hour of its birth. In this book I would like to show a way out of this endless debate. To achieve this, I have had to start off as a homeopath, then question my own methods. I started with the most frequent objections made by homeopathy critics and investigated these concrete points of criticism:

- From potency 6X (6D) onwards, hardly any active substance can be detected in homeopathic medication that could be responsible for physiological effects. (The dilution is too high at 1:1,000,000 to find enough homeopathic stock in it.) All homeopathic medicines with higher potencies are definitely free of active ingredients that could have a physiological effect.
- No "spiritual healing energy", as it was called by Hahnemann, can be proven in any grade of homeopathic potency.
- There is no "vital force" in the way Hahnemann imagined it.
- There is no principle of similarity in nature.
- From a scientific point of view, homeopathic drug testing is implausible and untenable.
- There are no studies that prove the effects of homeopathy beyond reasonable doubt; at best, an unspecific placebo effect can occur.
- The theory of homeopathy taken as a whole has no scientific basis and therefore cannot be part of today's medicine.

I also believed for a long time that the problem must be located on the side of science. Maybe scientists had not yet developed or discovered the right methods and standards to judge homeopathy through a full understanding of its effects. Stubborn scientists just didn't want to see any truth in homeopathy. The same held true for pharmaceutical companies, and because of this, they boycotted the appropriate research results. I long cherished the hope that science would not have the last word here. After all, as Shakespeare's Hamlet reminds us, there are many things between heaven and earth that cannot be explained scientifically. Every study can be manipulated, and according to the dictum "I only believe in statistics I have forged myself", scientific medicine cannot do better here either. In fact, that's what I thought until I began to deal with science myself.

One thing is certain though. No study meeting modern scientific criteria has been able to prove that homeopathy actually has any effect that goes beyond placebo (Ernst 2002; Shang et al. 2005), and its principles cannot be explained scientifically (or even made comprehensible).

That's what motivated me to get to the bottom of this, because I didn't want to offer my patients something that might be so vulnerable to criticism and sometimes even untrue. Patients must be able to trust that what we offer them is safe and correct. Of course, as patients you should not need to deal with scientific evidence. It's up to us, the therapists, to do that. In the following chapters, I will therefore deal with the points that have been shown to be wrong or highly vulnerable in homeopathy according to the scientific approach.

3.3 Spirit-Like Energy and Lack of Active Ingredient - The Problem of Potentiated Drugs in Homeopathy

In the natural sciences, *energy* is an established term with a clearly defined meaning, namely "the ability to perform work". Consequently, what is referred to as "energy" in homeopathy is not to be equated with the term "energy" as defined in science. Despite all the research in medicine and biology, no physical manifestation of such a power has yet been found. There has been no detection of a specific energy or spirit in any homeopathic medicine.

The frequently used term "subtle energy" is also plain and simply wrong. The proportion of active material is too low beyond a 6X potency to be responsible for any effect (Lambeck 2005; Hopff 1991). And there is nothing else in it that could be called "energy" or "spirit-like power". To be fair to Hahnemann, it must be noticed that most of the physical laws and the fundamentals of chemistry and physiology, as well as the possibility of molecular detection, were not yet known at his time. Therefore, he never had any way to really check his theories, even if he had wanted to. This makes a decisive difference between our present position and the way things were done then.

Another key point is that homeopathic medicines are supposed to have an "energetic" effect, or a so-called "spirit-power", or carry something like "information". Hahnemann attributed this peculiarity to homeopathic drugs because he assumed that the process of dynamization or potentiation could lead to such a phenomenon. The material parts are supposedly shaken away and the spiritual-energetic parts are shaken free. The higher the level of potentiation, the less physical material there is, but the more "dynamic power" should be contained in the medicine. I would like to quote an anonymous online article here because it is representative of many similar statements in which homeopaths explain the effect of their medication. I have deliberately chosen a very commonly used text:

> The beads are actually called globules. They are subject to a special method of production called potentiation. For example, a small bottle of Atropa Belladonna 12X contains as much homeopathic stock Belladonna (deadly nightshade) as if you had put a drop of it into Lake Constance, stirred

thoroughly, and put the water of Lake Constance into small bottles. Arithmetically, Belladonna 12X therefore means that one drop of the homeopathic stock is mixed with 1 million times 1 million drops of diluent. However, this would only be a very ordinary dilution. This could by no means show the kind of effect that the homeopathic remedy Belladonna 12X is actually able to do. In homeopathy, a remedy is not simply diluted or watered down, but rather potentiated. What does this mean? We mix 1 drop of the homeopathic stock with 9 drops of alcohol. This mixture is processed by hand by applying 10 shaking beats on a springy base to obtain Belladonna 1X. A drop is taken from this mixture and mixed once again with 9 drops of alcohol, and again 10 shaking beats are applied. That's how we get Belladonna 2X. According to this principle, all potencies are produced in the form of globules, drops, tablets, or powders. You may argue that in the high potencies there can't be anything left of the homeopathic stock! But it is sufficient that the information contained in the homeopathic stock should be included. (...) Due to the potentiation of the drug, some of the essential properties of the homeopathic stock are transferred to the diluent. In our example, the plant Belladonna is no longer present as a biological substance. However, it transfers its properties to a substantial medium (alcohol or lactose). When we listen to music on a cassette, it is clear that we don't have the whole orchestra in the cassette recorder. This is because it is possible to convert the information contained in the music into electronic vibrations which can be stored and played on a tape. Potentiation is similar to this.[1]

But even real (professional) homeopaths don't offer any other explanation. Irene Schlingensiepen's book "Homeopathy for Skeptics" (Homöopathie für Skeptiker) tells us about the oft-used term "information in the globules" (which is said is to be created by potentiation):

Suppose we order a dog to lie down. Let's assume now, in an excess of optimism, that the dog trots over to its basket, sighs, and climbs in. Of course, sound waves have transported the words "lie down" from the human mouth to the dog's ear. But it was not the effect of the sound waves that caused the dog to lie down in its place. After all, the sound waves didn't lift him up and carry him across the room. They only passed on the information to him, while it is the coded, molecule-free information that has persuaded the living organism to obey us. It is also possible to teach the dog to follow a pointing finger; the decisive factor is not which carrier we choose, but what information we transfer with it. (Schlingensiepen and Brysch 2014, p. 46)

---

[1]Source: Forum of rural women (in German), www.agrar.de/landfrauen/forum.

I have to say, this shocked me. The first explanation may perhaps be accepted as a simplification for the layperson. But that academic doctors claim that the effect of globules has been proven because one can tell a dog to lie down could hardly be less scientific.

In actual fact, I'm afraid most homeopaths and patients don't know exactly what potentiation really means. Maybe they envisage something extraordinary, but what really happens?

The homeopathic stock is triturated or treated with an extracting agent and solvent, diluted in the ratio described above, and then shaken vigorously. The solution contains the diluted homeopathic stock, solvents (e.g., water molecules, alcohol molecules), and impurities (e.g., suspended solids, dust, pollen, and other admixtures in traces). However, it remains unexplained (Lambeck 2005):

- Why only the desired substance, namely the original substance (the homeopathic stock), can be potentiated and not the rest.
- How a dilution and shaking process can generate energy from matter?

Hahnemann postulated that this would happen but failed to provide any explanation. And we as homeopaths still owe an explanation for this today. We are left with a dilution of a mixture containing, *among other things*, the homeopathic stock. This mixture is diluted further and further with increasing potentiation steps. Thus, it becomes less and less concentrated. In the higher potencies, the homeopathic stock is no longer detectable. The word "subtle" (or "ethereal"), which is often used to describe the specificity of homeopathic medicines, should therefore be more honestly replaced by "matter-free". In homeopathic circles, a higher potentiation means that a medicine is more potent (more effective) and contains more specific information, but from a scientific point of view this is incomprehensible. Physically and chemically it is not possible to justify the claim that potentiation differs from dilution. There is no possible scientific explanation for the idea that a transition can be obtained from an initially material active substance to energetic information by shaking. But even with normal common sense, there is no reason why some kind of spiritual energy should suddenly appear in a *shaken* diluted solution.

Neither can speculation at the quantum level that ultimately "everything is energy" (if that is actually right and understood by anyone other than a quantum physicist) explain what is supposed to have happened in such a shaken solution other than simple dilution. Even if the transition from matter to energy is a theoretically possible quantum-physical concept, it is not possible to use it to explain why something other than dilution should have occurred in the preparation of a homeopathic medicine or why such a transition should have taken place (Aust 2013; Lambeck 2005).

From a scientific perspective, I had even more difficulty understanding the theory of water memory or information storage in so-called water clusters. I could not regard this as proof of such an energy phenomenon either. In addition, there is a fact stated so succinctly and correctly by Aust: "The discussion about the retentiveness of water, which has been vehemently argued out in the controversy over the effectiveness of homeopathy, is also meaningless. The globules contain nothing of the water from the potentiation process except for the evaporation residues." Whatever happens in the potentiated solution, it is subsequently sprayed onto the sugar beads and evaporates. Of the "nothing" in the dilution will be leftover "nothing" in the globules.

It is not known for every homeopathic substance what concentration it should have in order to trigger a physiological effect on the body. To my knowledge, this has not yet been investigated. However, it is certainly no longer possible to assume a material effect with potencies above a 6X. Even at lower potencies, the concentration of active substance would not be sufficient (unless someone were to take the globules by the kilo).

In fact, as regrettable it is for a homeopath such as myself:

> The active ingredient of a potency 6X is so dilute that it can most likely no longer be responsible for any medicinal effect.

Just as the dose makes the poison, the opposite also applies, namely that "something" must be there to cause some kind of consequence. From a dilution or potentiation level 6X (i.e. 1:1,000,000, in words: one in a million), the dilution is definitely too high for a homeopathic drug to have any material effect associated with the original substance. In homeopathy, potencies 30C and above are referred as high potencies. In this case, the dilution has reached such a high degree that, with absolute certainty, no material effect is to be expected from the original substance (the homeopathic stock).

Hahnemann's idea, however, was that the material presence of a substance would not be necessary at all, for the "spirit-like" content was precisely what was intended to heal the spiritual vital force in the human body.

I had adhered to this idea for years, too. And I also prescribed high potencies for years. The idea is appealing enough. There is a life force in a person, an energy. Then a disturbance occurs, and this leads to an imbalance of the vital energy. This is expressed in externally recognizable symptoms. If we now approach the problem from a homeopathic standpoint, we describe the condition by capturing all external visible aspects of the disorder. We prescribe a drug that is very similar to this pattern of disorder, and, lo and behold, everything balances out. The drug must - and should - not be material, since we are dealing with a spirit-like problem. We have achieved this spiritual state of the drug through potentization. This seems internally consistent and easy to believe. The problem with such an energy is that it contradicts all known scientific principles. And since homeopathy claims to be medicine, i.e., science, and neither a religion, nor an agreement of faith or fantasy, it must be measured by the standards of science.

> What I want to absolutely clear about here is that no spirit-like energy or information, nor any inner spirit or universal essence, is extracted from the original substance (homeopathic stock) in any potency.

Shaking may be useful for the purposes of homogenization, but it won't under any circumstances provide spiritual energy or information.

I've read many texts in which explanations such as "water memory", quantum physics, and other ideas have nevertheless been used to prove or explain the effectiveness of immaterial potencies. Unfortunately, no one has ever convinced me. And I really would have liked to be convinced! The referenced literature for this chapter contains even more sound arguments and I recommend that you take a look at these sources (Aust 2013; Ernst 2013, Lambeck 2005; Shang et al. 2005).

The explanation with the music cassette quoted above, where one no longer has any material information after all, does not stand up to closer scrutiny. In contrast with the globule, the information is clear and it is precisely known how the information is stored on the cassette, in such a way that this process contradicts no scientific principle. No cassette manufacturer would claim that some kind of spirit would be stored on his medium, after transferring itself from the musicians to the tape! The same applies to the dog which has learned to obey an order: the relations causing it to follow our instructions are certainly complex, but they can be explained, and they can be explained without reference to some kind of mysterious information energy.

These homeopathic explanations remind me of the phlogiston theory in the 18th century. At that time, it was hard to understand why less material was left after combustion than had originally been present. The term "phlogiston" was coined to describe a previously unrecognizable substance, a hypothetical component that was assumed to escape from all flammable substances during combustion. This theory was later refuted when it was discovered that some substances burn to produce gases. Phlogiston was thus an idea, a conjecture, that was refuted by advancing scientific knowledge. I fear that this is now also the case for Hahnemann's homeopathy. Hahnemann's ideas are exactly that: edifices built up from conjectures. Outdated ideas that we can no longer retain. Those who adhere to this construction without accepting the further development of science are, so to speak, becoming "devotees of phlogiston". This has nothing to do with science. And, in consequence, nothing to do with medicine.

In our time and with the present state of scientific knowledge, Hahnemann's ideas are no longer acceptable - at least not as facts on an equal footing with those of science. It is never easy to frame a universal law of science, such as a physical law. It must be consistent, comprehensible, and complete (Lambeck 2005). Let's assume that physicists understand their business and that the laws that have been established up to now meet these stringent demands. If homeopathy wants to be part of the natural sciences, it must not contradict these laws and the consistent structure they provide. This also applies to the content and meaning of individual terms such as "energy", "information", or "force".

3.4 The Problematic Concept of Vital Force

A further problem in homeopathy as medicine is that Hahnemann postulates a kind of vital force within the human being, which is responsible for illness and which must be influenced to restore health (by the potentiated drugs). The term "vital force" is one of the most basic concepts of homeopathy. In paragraph 9 of his Organon, Hahnemann defines it in this way for the first time:

> In the healthy condition of man, the spiritual vital force (autocracy), the dynamis that animates the material body (organism), rules with unbounded sway, and retains all the parts of the organism in admirable, harmonious, vital operation, as regards both sensations and functions,...
> Hahnemann (1997), Organon, Paragraph 9

He continues in Paragraph 10:

> The material organism, without the vital force, is capable of no sensation, no function, no self-preservation, it derives all sensation and performs all the functions of life solely by means of the immaterial being (the vital principle)

which animates the material organism in health and in disease. Hahnemann (1997), Organon, Paragraph 10

In Paragraph 12 he says:

> It is the morbidly affected vital energy alone that produces disease, so that the morbid phenomena perceptible to our senses express at the same time all the internal change, that is to say, the whole morbid derangement of the internal dynamis; in a word, they reveal the whole disease. Hahnemann (1997), Organon, Paragraph 12

And in paragraph 16 he describes how the disorder of the vital force can be retuned:

> Our vital force, as a spirit-like dynamis, cannot be attacked and affected by injurious influences on the healthy organism caused by the external inimical forces that disturb the harmonious play of life, otherwise than in a spirit-like (dynamic) way, and in like manner, all such morbid derangements (diseases) cannot be removed from it by the physician in any other way than by the spirit-like (dynamic, virtual) alterative powers of the serviceable medicines acting upon our spirit-like vital force, which perceives them through the medium of the sentient faculty of the nerves everywhere present in the organism, so that it is only by their dynamic action on the vital force that remedies are able to re-establish and do actually re-establish health and vital harmony. Hahnemann (1997), Organon, Paragraph 16

In his idea of vital force, Hahnemann followed the ideas of his time. His contemporary and medical colleague Hufeland, for example, considered the basic cause of all life processes and the self-preservation principle of the organism to be a general vital force with further partial forces:

- a sustaining force,
- a regenerating and rebuilding force,
- a special life force of blood,
- a force of nerves,
- a force that effects a *general* irritability of the body, and
- a force that effects a *specific* irritability of the body.

If Hahnemann postulated a kind of spiritual vital force that would be disturbed by illness, this was quite adequate to the science *of his time*. In Hahnemann's day, it was not unusual to think like that. The doctor's aim was to obtain an exact picture of the disorder in the assumed vital force. But

Hahnemann went even further than his contemporaries: it should also be the doctor's aim to find a drug that was similar to this disorder, which was supposed to release the vital force and get it back into free flow. At that time such an idea may have seemed adequate - but what about today? The problem is that neither physics, nor chemistry, nor medico-physiology, nor anything in modern science knows of any such vital force inherent in the body. I will therefore think of the term only as follows: since such a life force cannot be represented scientifically, it can exist only as an idea, a figment of imagination in homeopathy. It is a concept that was better suited to the early 19th century, when such ideas stood in for otherwise lacking scientific insights. But now that we have developed the natural sciences so significantly, there is no place for such a concept. Please understand my purpose here. It is not my intention to replace one term (vital force) with another (spiritual idea, figment of imagination). Nothing would be gained by that. I want to make it clear that Hahnemann's notion of life force was a mere idea, a pure product of the imagination. Such conjectures were used to explain the unexplainable at the time. Today we must no longer adopt the term in medicine - at least not as a fact.[2] I therefore propose to regard the vital force in our context as a personal idea put forward by Hahnemann.

Imagination and ideas are good things, unless they are confused with science and facts. But we must be clear: despite all the research in this field, no physical or biological instance of this homeopathic idea has ever been identified.

Complex phenomena such as consciousness, imagination, and ideas are "more" through their interactions than just physics and biology. However, they can never lie outside of all physical possibilities - and indeed they do not - the basic physiological processes are known and have been described. Their "inexplicability" as phenomena is due to their immense complexity. They do not require the assumption of an imaginary "vital force" in order to gain an interpretation, because they do not contradict basic scientific knowledge. They do not lie outside the applicable standards. This is not the case with the concept of vital force in homeopathy. The *vis vitalis* (which also existed in other doctrines of salvation in Hahnemann's time and before) is thus refuted.

---

[2]For explanation, it should be added that this applies to the natural sciences and not every science. In the social sciences, for example, such idea-based constructs are used when considering phenomena to explain the theory. These constructs are nothing else but ideas with explanatory potential. Phenomena ascertainable in the social sciences do not necessarily have a material substrate and can also involve ideas, feelings, or sensations. The social sciences (e.g., psychology) have a different research subject than the natural sciences, and only natural science has so far been decisive for medicine. That is why I am referring here to the latter.

So today, in good conscience we can no longer treat the vital force of homeopathy as a fact. However, we can see that Hahnemann's idea of a life force may not be so far removed from the ideas of some of our patients today. The term is still known and used today as a colloquial description for similar views. Let me summarize some of these ideas, which may be roughly described by the following points:

- subjective health experience
- regenerative capacity
- personal disposition and constitution
- vigor
- willingness to assert oneself
- desire for life
- healthy development of one's own potential
- spiritual idea of something superordinate in man
- heritage, perhaps described nowadays by the term "genetic endowment".

I still consider it a great asset of homeopathy to deal with these issues and to take such far-reaching points into consideration. In this respect, we must not always take Hahnemann's words literally, even though we may perhaps continue to accept them as ideas. Many patients are spontaneously open to the idea of life force and connect individually and intuitively with some of the points mentioned above. The idea of turning such a vital force into a harmonious flow by recognizing where it has been blocked seems to attract and even inspire many patients, and maybe this is what still draws them towards homeopathy.

3.5 Homeopathic Drug Testing (Homeopathic Pathogenetic Trials - HPT)

In the chapter on homeopathic diagnosis, I already mentioned how homeopathic drug testing is carried out. In short, a healthy person takes the drug to be tested, and the changes in the physical and emotional realm, but also in the realm of sensations, are recorded and categorized according to frequency, etc.

When I first perused the description and the result of such an "examination" as part of the research for this book, I seriously wondered how homeopathy could be built on such a procedure. I had simply not dealt with it before (as, I am sure, many homeopaths and patients have not) and was appalled at what was supposed to be the basis of our homeopathy. All the symptoms that had occurred in the subjects seemed to be randomly mixed up: dream symptoms as well as colour sensations or special self-perceptions, but also completely contradictory physical symptoms, such as pain worse due to pressure/better due to pressure). As a homeopath I know about *Boenninghausen's* valence analysis[3] and that such contradictory symptoms can be weighted. But even if we leave this problem aside - how do we explain all this when we consider that it is supposed to be caused by drugs that do not contain anything?

Put briefly, drug testing in scientific medicine proceeds as follows. First, a meaningful hypothesis on the mechanism of action of a drug must be available and the cause of a disease must be clear. Then it is examined whether the proposed drug (together with its assumed mechanism of action) leads to a repeatable effect. In addition, it is also tested whether such an effect does not occur if the drug is *not administered*.

> In scientific drug testing, the cause-effect relationship is decisive and the rule in modern medicine. It is not only important that A occurs when B is applied, but also and above all that A does *not* occur when B is *not* applied.

This means for homeopathy: one gives A, and B (test symptoms) occurs. But how can we be sure that B was caused by A? Homeopaths, including Hahnemann, simply assert this, and rely on it to justify use of the drug. But is it true? Further questions arise, which I am sure I am not alone in asking:

- How do we know for certain that the examinees are "healthy", i.e., that they do not slip in any symptoms of their own possibly imperfect condition, especially symptoms of the emotional state?
- How can we be sure that it is the homeopathic medication that caused the symptoms?
- How can we check this? (There are now blind drug trials where examinees do not know whether they are receiving a placebo or a real homeopathic drug, but what could that do for us if there is nothing in either that can be responsible for an effect?)

---

[3] *Clemens Maria Franz Freiherr (Baron) von Boenninghausen* (1785–1864) was a lawyer, botanist, physician, and pioneer of homeopathy. His "Therapeutic Pocketbook" of 1846 was the first homeopathic repertory to grade individual remedies by their strength of relationship with each symptom, and with each other.

- Could Hahnemann's cinchona self-experiment have been the first false test? Who can be certain that he was healthy, that there were no other influences, or that everything was not simply a coincidence? After all, the experiment could never be reproduced: nobody else ever showed the kind of symptoms Hahnemann described after taking cinchona bark!
- Have we homeopaths repeatedly made the same mistake since then, administering placebos, writing down all the changes we notice after they are taken, and maintaining that there must be a causal connection?

Some modern homeopaths take a completely different approach (e.g., Scholten, Sankaran): they have preconceptions. They assume what effects the selected drugs should have according to the systematics of their own teachings and make trials on this with healthy persons under these assumptions. For example, the sensation of *Sodium muriaticum* (table salt) or *Rhus toxicodendron* (poison ivy) are clear for Sankaran. From the always numerous symptoms revealed in the trials, the confirming symptoms and sensations are now adopted - and behold, it fits...

I can only invite every patient and homeopath to take a closer look at the theory, but also at the practice of homeopathic drug testing. I would never previously have believed what is happening here. From psychological studies, we know that what we expect is more likely to happen than the unexpected (Wikipedia, keyword List of cognitive biases; Herrmann 2013). What homeopaths are delivering here is a self-fulfilling prophecy - not medicine in today's sense.

I won't deny that repertories often contain homeopathic drug images that show us very complex and differentiated personality studies. And I can also confirm that I already had patients in my practice who were astonishingly similar to these descriptions - so I also *felt* this gave credence to what I was doing myself. However, we cannot use them as actual confirmation of our homeopathic theory of similarity or as a proof of potential efficacy. If we did so, we would draw hasty and most likely erroneous conclusions based on analogy, whereas what matters is slow, arduous, scientific thinking: what errors might we be making by thinking in this way? What other influences might there be? How could we prove that A has led to B? Hahnemann was not familiar with this way of thinking, so we can't blame him. Today, however, we must recognise and admit that homeopathic drug trials are anything but a trial. And with this, the principle of similarity is untenable. If the tests result in only random constellations, we can't use them to create reliable drug images that we might compare with patient images.

3.6   Is Homeopathy Medicine?

It was possible for homeopathy to be a part of medicine in Hahnemann's times. But today, medicine is much more advanced and has changed significantly. Research in physics, chemistry, biology, biochemistry, physiology, and pathophysiology, but also in scientific theory and statistics, have led to insights and fundamentals that just did not exist at that time.

Medicine is the attempt to apply these scientific findings plausibly for curative purposes and to test this procedure - either subsequently or in advance on a theoretical basis - for causal connections and significant results (evidence-based research). It is true that shortcomings in the implementation of the findings of such research also become apparent in scientific medicine. I don't want to go into these here, as this book is dedicated to other issues. However, the principles of natural science and medical research are clear and generally accepted. Of course, every science must find criteria that are relevant to its research subject and relative to which its claims can be tested. Archaeology differs from theoretical physics, economics from engineering, and so on. If homeopathy wants to be part of medicine today, it must adhere to the principles developed within medicine or develop its own comprehensible criteria, for example analogous to those of psychology and other social sciences. But at the present time, it just does not do so, or only to a very limited extent. Critics are therefore right when they call on homeopathy to this.

The problem is this. On the one hand, we can't claim that homeopathy works and still has a justification in medicine today, but on the other hand, to justify this claim we can't start from principles that don't correspond in any way to those of the natural sciences. Just doing more (doubtful) studies won't help to close this gap. We won't be able to prove anything in this way. Neither the production, nor the testing, nor the common use of our homeopathic medicines are scientifically comprehensible and verifiable. I will deal later with a special side effect of administering globules, the placebo effect, which does belong to medicine.

The concepts of vital force, energy, spirit, etc., can't go unquestioned as if it is established that they refer to facts.

As a result of these considerations, homeopathy can clearly not be considered part of today's medicine.

So why waste any more time on homeopathy? Concepts from times long past, strange ideas about drug effects, esoteric terms, tests that were never really tests - what is the point of that? Does homeopathy nevertheless have some influence on human health? An influence that isn't in conflict with science? Why do so many patients still turn to it? These are the questions I would like to address in the next chapter.

References

Aust N (2013) In Sachen Homöopathie – eine Beweisaufnahme. (The case of homeopathy - the evidence) 1–2-Buch, Ebersdorf (in German)

Dobelli R (2014) The art of thinking clearly: better thinking, better decisions. Sceptre, London

Ernst E (2002) A systematic review of systematic reviews of homeopathy. Br J Clin Pharmacol 54(6):577–582

Ernst E (2013) Natürlich heilen. Gesund mit sanfter Medizin, Heft 2013/4. www. spiegel.de/spiegelwissen/alternative-heilmethoden-edzard-ernst-ueber-die-wirkung-vonglobuli-a-934517.html. Accessed 22 April 2018 (in German)

Hahnemann S (1997) Organon of medicine, 6th edn (Translation Boericke). http://www. homeopathyhome.com/reference/organon/organon.html. Accessed 20 April 2018

Herrmann S (2013) Starrköpfe überzeugen. (How to convince stubborn minds) Rowohlt, Reinbek bei Hamburg. (in German)

Hopff W (1991) Homöopathie kritisch betrachtet. (Homeopathy viewed critically) Thieme, Stuttgart (in German)

Kahneman D (2012) Thinking, fast and slow. Penguin, reprint edition

Lambeck M (2005) Irrt die Physik? (Is physics wrong?), 2nd edn. Beck, München (in German)

Schlingensiepen I, Brysch MA (2014) Homöopathie für Skeptiker. Wie sie wirkt, warum sie heilt, was belegt ist. (Homeopathy for sceptics. How it works, why it heals, what is proven) Barth, München

Shang A, Egger M, et al (2005) Are the clinical effects of homeopathy placebo effects? Comparative study of placebo-controlled trials of homeopathy and allopathy. Lancet 366(9487):726–732

4

Why Do Patients Turn to Homeopathy?

It's strange to observe that homeopathy is still very much in demand despite the considerable shortcomings that have been pointed out. In August 2014 Amazon listed 17,107 books on homeopathy. This includes some textbooks. But many books on homeopathy are just pure DIY guides, including some bestsellers. Many magazines contain health tips based on homeopathy.

According to a study carried out by the German institute for public opinion research Allensbach in 2009, the proportion of the (West German) population (16 years and older) who have already consciously used homeopathic medicines has increased considerably over the last forty years. Whereas in 1970, 24% of the respondents had themselves used homeopathy, in 2009 the figure had increased to 57%. Still in 1970, 32% had never previously heard of homeopathy, while in 2009 this proportion had decreased to 6% of the respondents. In October 2014, the Allensbach Institute once again surveyed many citizens on behalf of the German Association of Pharmaceutical Manufacturers (Bundesverband der Arzneimittelhersteller, BAH): more than half of the population in the sample had already used homeopathic medicines, and compared to 2009, the proportion of users rose further to 60%, with 73% of the users being women. Nearly nine out of ten users reported that homeopathic medicines had helped them, and 48% said so without any restriction. A further 39% reported that homeopathic medicines had helped them, at least in some cases (BAH, press release of 20.10.2014[1]).

---

[1]https://www.bah-bonn.de/presse/pressemitteilungen/artikel/repraesentative-befragung-immer-mehr-menschen-nehmen-homoeopathika/.

© Springer Nature Switzerland AG 2019
N. Grams, *Homeopathy Reconsidered*,
https://doi.org/10.1007/978-3-030-00509-2_4

Demand and interest on the part of patients are obviously high. Alternative medicine continues to boom and homeopathy plays an important role in this. In many cases, patients have already consulted their family doctor or a specialist. They often came to me with large files of findings, or reported on psychotherapy in their case history. So why were they coming? What makes homeopathy so attractive for them, especially since they must know at least some of the criticisms that can be made of homeopathy? Why does everyone you talk to about homeopathy know at least one friend it seems to have helped, or have their own story to tell? Are all these narratives just anecdotes, or can homeopathy actually do something that is worth following up - despite all the criticism?

Debates in the many online discussion forums on homeopathy, interviews with my patients, and conversations with other homeopaths can give us a list of reasons why patients turn to homeopathy:

- it is a "holistic" approach;
- it treats feelings, mind and soul;
- there is scope for more spiritual questions;
- there is scope to take into account emotional hardships;
- one is perceived as an entire human being;
- the problem is not reduced to a single symptom;
- the homeopath has time for the patient;
- the homeopath provides more exclusive attention;
- "regular" medication is avoided when it is not absolutely necessary;
- it seems to be an alternative to scientific medicine;
- it seems to offer a more active role in personal healthcare;
- it is a talking (or listening) medicine;
- it provides understanding of rather unusual worries, going beyond the symptom alone;
- it provides a second opinion;
- it provides life guidance;
- it is a response to bad experiences with the normal health care system (e.g., little time, fear of side effects with normal medication, etc.);
- it is often a kind of faith, or almost a dogma, that leads patients to homeopathy.

How can homeopathy respond to these wishes? What does it offer that attracts patients? And how exactly does it achieve that?

Hardly any opponent of homeopathy doubts that homeopathy has an effect. I too can confirm this from my daily practice: homeopathy helps some people, it does them good. In the following sections, I'm going to refer to the advantages of homeopathy *as a method* and ask whether it is possible to concretize the "good feelings" of the patients. This is undoubtedly important for the effects indicated by the patient. I will deal with the aspects that are important from the patient's (and partly also from the therapist's) point of view in the actual practice of homeopathy and I will return to the scientific processing of these "subjective" points of view in the next chapter.

4.1   The Therapeutic Setting of Homeopathy

4.1.1   Time, Empathy, and Care

We are probably all familiar with this. With an acute or chronic problem, we sit in the doctor's waiting room or in the clinic. After a more or less lengthy wait, the door opens, we hurry into the treatment room - and we are out again five minutes later, carrying a prescription or a referral to a specialist. I remember my time in the clinic when a senior physician gave the following instructions: "The ward round is not a colloquium for discussions or exchange of ideas. It depends on rushing forward and then immediately organizing the retreat." Surely this was meant partly in jest, and the senior physician was a very good surgeon, but there was some truth in it. In my time as a ward physician, half a minute at the patient's bedside was a lot. Sometimes we even had to pare down that small amount of time to get the job done at all. This was also true at the surgeries I worked in, where one patient followed another, and it was difficult to grant more time for any given patient while in the meantime the waiting room continued to fill up. In modern medicine, time is an almost non-existent commodity.

In homeopathy, however, there is plenty of time. Hahnemann expressly insisted that his students should let patients finish speaking, no matter how long this might take, and then inquire again until they could be sure that everything really had been said. In paragraph 84, for example, Hahnemann writes about this in his organon:

> ... the physician sees, hears, and remarks by his other senses what there is of an altered or unusual character about him. He writes down accurately all that the patient and his friends have told him in the very expressions used by them. Keeping silence himself he allows them to say all they have to say, and refrains from interrupting them unless they wander off to other matters. ...
> Hahnemann (1997), Organon, Paragraph 84

An initial professionally conducted homeopathic consultation usually lasts between one and three hours (I know colleagues who even spend up to six hours with a patient) and a follow-up interview lasts half an hour to one hour. Only in psychotherapy is a similar amount of time available. This time factor is decisive for many patients. Homeopathy is often referred to as the "talking medicine"- this is a great advantage offered by homeopathic treatment. From the therapist's point of view it is a "listening medicine"; the homeopath must be able to listen until patients have really said everything that is important to them, and this without judging them. It is obvious that such attention will be of benefit to the patient. Moreover, if this listening is done with an empathic attitude, i.e., with openness and sensitivity, it will certainly be even more effective. The patient feels accepted and understood, not judged or - in the worst case - simply fobbed off. Given how little time we have in our everyday lives, such a gift of time as can be expected in a homeopathic consultation is appreciated even more. Exclusive attention is certainly one of the great advantages of homeopathy, which its critics would not deny. Time in homeopathy also means being able to contact your therapist at any time. Most homeopaths offer fixed telephone consultation hours and can also be reached at short notice by e-mail or for telephone inquiries. Patients receive advice and support beyond what is possible for the general practitioner. Therapeutic monitoring is also more intensive in this respect.

Basically, homeopathy adopts the method of "active listening" known from psychology. The goals of active listening are multi-layered. On the inter-personal level, especially on the relationship level, the aim is to build up mutual trust and promote an appreciative form of interaction. On the semantic level, active listening helps to avoid misunderstandings. It is intended to improve problem solving, to make simple behavioural corrections, and to enable learning through feedback.

Certainly one of the significant goals of any counseling experience is to bring into the open those thoughts and attitudes, those feelings and emotionally charged impulses, which center around the problems and conflicts of the individual. [...] Consequently the counselor must be skilled indeed in providing release for the client in order to bring about an adequate expression of the basic issues in his situation. (Rogers 1942)

4.1.1.1   Criticism of This Aspect of Homeopathy

This particular homeopathic procedure generally receives little criticism. In fact, quite the opposite: the possible positive effects are attributed above all to this procedure.

4.1.2 Individual Point of View

Many patients hope their symptom or disease will not just be picked up in the form of a diagnosis, behind which they will promptly disappear as a person. Instead, they want their personal views and individual feelings to be taken seriously and included somehow in the treatment. This procedure does not correspond to what happens in normal medicine. In homeopathy, on the other hand, it is precisely this desire that will be satisfied. From a homeopathic point of view, a symptom is an indication of the patient's particular disorder. A symptom must always be seen in the context of a patient's individuality, and the individual patient picture emerges from the symptom determination. A symptom never stands by itself, and nor is it sufficient to make a diagnosis.

According to Hahnemann,

> ... the more striking, singular, uncommon and peculiar (characteristic) signs and symptoms of the case of disease are chiefly and most solely to be kept in view ... Hahnemann (1997), Organon, Paragraph 153

In homeopathy, therefore, it is not possible to prescribe a medicine against back pain without having spoken to the patient personally or at least having obtained a precise description of the pain (e.g., from the parents of an affected child). In contrast to the view of scientific medicine, the patient is more competent than the therapist in identifying the special and "peculiar", i.e., specific symptoms. Only the patient knows how to describe the symptoms because he or she feels them in a totally unique way. This attribution of competences certainly has its own therapeutic effect. There is no homeopathic medicine against back pain; rather, there is an individual patient picture that needs to be elaborated. Homeopathy is highly individualised - in regard to both medical history (anamnesis) and the prescription of medication.

Individualisation can also be just as important in self-treatment. The fact that I don't just take *any* drug against pain, but in fact one designed to help me with *my* pain (which, as described in the initial case, for example can be severe, constricting, stiffening, initially worse when moving, then better, etc.) seems to have an additional positive effect on recovery, simply because my own sensations are taken seriously and emphasized.

Another advantage of the individual approach is to place symptoms in a context that makes sense to the patient. One patient reports: "I felt so trapped at work - and that led to my symptoms!" With homeopathy we can take such an unusual statement seriously and follow it up.

However, Hahnemann also makes it clear that not every little ailment should be followed up and treated:

> If a patient complains of one or more trivial symptoms, that have been only observed a short time previously, the physician should not regard this as a fully developed disease but requires serious medical aid. A slight alteration in the diet and regimen will usually suffice to dispel such an indisposition. Hahnemann (1997), Organon, Paragraph 150

However, it will once again take time to differentiate between largely irrelevant minor coincidences on the one hand and actual specific symptoms on the other in each individual case!

4.1.2.1 Criticism of This Aspect of Homeopathy

The individual approach of homeopathy is not generally criticized. On the contrary, even in scientific medicine there is also an increasing consideration of individualisation, for example, regarding the dosage of medicines.

4.1.3 Being Able to Do Something

In conversations with doctors and paediatricians, I've often heard: "Being completely honest, we would often have to say to our patients: "Unfortunately, we can't do anything about that now. Only time, rest, and perhaps tender loving care can help."

This is mainly the case with rather trivial infections, but it may also be the right approach at times for serious diseases. Of course, there is always a strong desire to actually do something, not to wait around helplessly and passively. However, it may often be necessary to endure uncertainty and perhaps take time off from work, school, or kindergarten in order to keep still and rest. However, this is not always easy to do and it may seem unusual and frightening, for both patients and therapists: in particular, doctors may fear that they will be considered incompetent if they propose something like this - without taking any "real" action for their patients.

It is much better to get a prescription from your homeopath or to get some globules with a very precise prescription; or indeed to work out this medication yourself using a quickfinder and take globules every three hours. After a while the discomfort disappears - and we are convinced that it was due to the homeopathy. But could it not just have been the elapsed time? Aren't we - by "being able to do something" - just making it easier for ourselves to accept the time necessary for recovery?

Of course, it couldn't be the main aim of a method to keep the patient in a good mood until their body has overcome the disease by itself. But wanting to offer comfort and confidence and hope in an emergency, - something to hold onto, as it were - this desire is understandable, and it is indeed satisfied on the individual level by homeopathy.

In my practice I noticed that it does patients good if they can make a call and ask questions whenever there is an acute worsening of their condition or a new disease appears. This availability is part of the therapy. It helps to know that there is always someone you can talk to. This level of availability is usually not possible in normal medicine. And at least, such a phone call is the first step toward doing something.

4.1.3.1 Criticism of This Aspect of Homeopathy

Such a conversation-oriented procedure should in no way delay or neglect a necessary and meaningful conventional medical treatment. I'm fully aware that in this respect some criticism of the approach can be justified (Weymayr and Heißmann 2012). It is more appropriate in situations where wait-and-see is possible, due to the often mutual expectations of physicians and patients that "something must happen". I am referring, for example, to harmless viral infections, periods of convalescence, commonplace well-being disorders (which are often perceived completely differently by the patient), latency times up to the onset of a drug's effect, and palliative cases. From the patient's point of view, time can often be a valuable factor, and empathic support worth more than a pointlessly given medical remedy. Nevertheless, I am aware that there is often negligence here - and that this is not to the benefit of homeopathy.

Critics also complain about prescribing globules just for the purpose of "doing something", because "constant globalisation" would accustom patients the assumption that a medicine is always needed and an essential factor in getting or staying healthy. The honest statement "We can't do anything now but wait and see" would in fact sometimes be more appropriate. I have heard paediatricians criticising the fact that mothers too often give globules to their sick or indisposed children. I would agree, I would also say that the use of a correct but unnecessary conventional medicine should by no means be the alternative.

This procedure is particularly risky if patients believe that homeopathy can *always* have some effect. Hahnemann has already taken a stand against this, for example by excluding surgical cases from homeopathy (Organon, paragraph 13). But honestly, I must admit that I know homeopathic colleagues who really demonise *all* drugs proposed by normal medicine, and instead swear by homeopathy - even in cases where it is clearly dangerous and irresponsible: the idea behind this is that everything can be cured through homeopathy, homeopathy can always do something. This is absolute nonsense and has nothing to do with the advantages of homeopathy, that is, remaining able to act as a patient and to do something for oneself, at least if there is no danger in delay.

4.1.4 No Side Effects

Another argument that patients often cite in justifying why they have chosen homeopathy is the claim that it does not cause any side effects. This is associated with the attributes "gentle" and "natural". Homeopaths and pharmacists support this assessment. Homeopathy does indeed offer patients the possibility of taking gentle medication that is free of side effects. In cases where there is really no need for conventional medicine, this may be acceptable. Many patients seem to see the globules as a harmless and side-effect-free alternative, with which they can "do something" (without actually taking anything). In such cases, patients may be perfectly aware that no proper treatment with medication is necessary, but still have a great desire for help. So if it is possible to take harmless globules, which are said to have no side effects, they will often choose this alternative.

Homeopathy offers therapists the opportunity to prescribe something without really giving anything - simply with the effect of having done something and avoiding normal medical drugs where those aren't strictly necessary, and without having to reckon with side effects.

The use of globules *against* the side effects of conventional medicine is also a popular practice. Many patients realize that normal medicine together with its possible side effects are inescapable when treating a serious disease (e.g., chemotherapy for cancer). They are often pleased to think that they can do something about the side effects by taking homeopathic globules in these cases. Scientific medicine probably underestimates people's fear of the side effects of normal drugs. I would spend a lot of time in my practice trying to allay such fears, or at least paying attention to them. Compliance (reliability of actually taking the medication) decreases under this fear. For example, many of my patients told me that they do not ingest their blood pressure medication according to the doctor's instructions because they are afraid of the side effects listed inside the packaging, or because they have already experienced these side effects and want to avoid them in the future. However, they do not tell their doctors this. They come to believe that they can tackle this problem with homeopathic medicines and are supported by homeopaths.

However, there are also side effects in homeopathy:

As a side effect, homeopaths see a temporary amplification of the symptoms, which they call homeopathic first worsening (also initial aggravation). (…) In the case of low potency levels (up to about 6X), an undesirable drug effect can occur, because the drug still contains appreciable amounts of substances. For example, poisoning can be caused by using of mercury, arsenic or Nux vomica (poison nut), a plant containing strychnine alkaloids. Wikipedia (German version), Keyword Homeopathy

Initial aggravation is a temporary worsening of symptoms. From a homeopathic point of view, it is caused by the artificial disease that the drug is supposed to trigger in the body. However, there is no evidence that initial aggravation actually exists (Grabia and Ernst 2003).

The advantage is that a - real - worsening of the symptoms is perceived as reasonable and can be tolerated better until natural spontaneous healing occurs. In this respect, however, homeopathy is not free of side effects.

4.1.4.1   Criticism of This Aspect of Homeopathy

Critics complain that the worst side effect of homeopathy is that it gives a promise of salvation that it is unable to fulfill, and in the worst case this can lead to the omission of absolutely necessary medical intervention. This is unfortunately true when homeopathy is practised in cases where conventional medicine is clearly needed. Due to the expectation of the frequently predicted initial worsening, it may become dangerous if the patient is induced to wait and no adequate medical help is provided, even after a certain time has elapsed (Weymayr and Heißmann 2012).

4.1.5 Deep Doctor-Patient Relationship

The merits of the homeopathic therapeutic setting described above lead to a further advantage: the exceedingly close doctor-patient relationship. In everyday clinical and practical medicine, the patient's contact with the doctor is limited to a few minutes and to a rather impersonal relationship.

Of course, doctors in conventional medicine are also subject to the medical ethos.

The Medical Ethos

- Respect for patient autonomy

The principle of autonomy concedes to everyone the freedom of choice and allows patients the right to make their own decisions. It includes the requirement of informed consent (...) before any diagnostic and therapeutic measures are taken, and the consideration of the patient's wishes, goals, and values.

- Non-maleficence (do not harm)

This is the principle that damage must be avoided. This may seem self-evident at first, but it often comes into conflict with the principle of care for certain therapies (e.g., chemotherapy).

- Welfare, assistance

The principle of welfare requires the therapist to act in a way which promotes the well-being of the patient and is useful for the person concerned. The welfare principle is often in conflict with the principles of autonomy and damage avoidance (...). Risks and benefits should be carefully weighed up, taking into account the wishes, objectives, and values of the patient.

- Equality and justice

The principle of justice demands a fair distribution of health care services. Equal cases should be treated in the same way, and in the event of differing treatment, morally relevant criteria are called for.
From: German Wikipedia (Keyword Ethics in medicine)

However, many patients have learned that doctors often fall far short of this theoretical ideal. When do we really respond to the patient's goals and values in everyday medicine, where time is in such shortage? Patients are indeed likely to learn that they are all being treated equally: in fact, as quickly as possible. So, it's not surprising that they turn to therapists who offer them their time, extensive listening, great understanding, individual therapy decisions, and even individually tailored medication.

A (first) consultation in homeopathy takes a long time and the therapist will spend this time exclusively with the patient. The ensuing procedure will also be closely coordinated between doctor and patient; in most cases, this means that the patient regularly returns to follow-up talks and that interim consultations are possible at short notice. Henceforth, the homeopath plays a major role in the patient's life. I know patients who travel from afar to meet with "their" homeopath, often with the whole family. Therefore, great hopes are often invested in the homeopath as a person - in my experience, far greater than they would be for a general practitioner.

It is clear that such a relationship will have an influence on the patient that is in itself equivalent to an independent therapeutic effect.

4.1.5.1 Criticism of This Aspect of Homeopathy

The intimacy in the relationship between homeopath and patient can quickly lead to a kind of dependency or even bondage. Some homeopaths profit either consciously or unconsciously from the patient's need for an alternative to scientific medicine or the search for human empathy. Homeopaths are often convinced that they are doing the right thing. This leads to legitimate criticism and malice against homeopathy, and works rather against the fulfilment of the medical ethos.

In psychotherapy, moreover, it is clearly excluded that the therapist should exert an excessive influence on the patient through his or her personality and charisma. But I have experienced this more often in homeopathy and it should be viewed extremely critically.

4.1.6 Holistic Approach

The term "holistic" is often used when speaking about homeopathy. The term was introduced in psychology in the first half of the 20th century. In the 1970s it found its way into the concept of psychosomatic medicine. It describes an approach that attempts to link several biological, psychological, and social factors in medical history and diagnostics (Wikipedia, keyword Holism, items Medicine and Psychology). At Hahnemann's time, however, the term was not yet part of the conceptual toolbox. But what does it really mean? Even during my studies I found that the term is taken to imply rather a lot, but often remains somewhat abstract. What does it mean to be seen and treated "holistically"? I am aware that the classic risk of esotericism lurks in such consideration of the "big picture". The qualitative leap from individual parts to the "whole thing" always potentially contains the possibility of something not completely scientifically explainable - and also of some blurriness and distortion in one's own point of view. I have therefore tried to examine the supposedly holistic aspects of homeopathy more closely and in more concrete terms (this also includes the following section on the homeopathic clinical picture).

In the German version of Wikipedia, we find the following good definition of holistic medicine, which I would like to quote here in translation:

> Holistic medicine is an approach to health care in which the whole person should be considered and treated in his or her life context with the emphasis on subjectivity and individuality. (...) According to this, each individual is a structured system that is open towards the outside and whose parts would be in mutual relationship with each other, with the whole organism, and with the outside world. Factors to be taken into account for medical treatment would therefore be the unity of body, soul, and mind, the patient's ideals and values, his or her lifestyle (movement, nutrition, stress, relaxation), the social environment including all human relationships (partner, family, profession, fellow human beings, society), the natural environment (water, soil, air, climate), the artificial environment (housing, workplace, technology) and, although not generally accepted, the supernatural (religion, belief, spirituality).

Anyone who has ever experienced a homeopathic anamnesis and treatment knows that these conversations can be very comprehensive:

> When the narrators have finished what they would say of their own accord, the physician then reverts to each particular symptom and elicits more precise information respecting it in the following manner; he reads over the symptoms as they were related to him one by one, and about each of them he inquires for further particulars, e.g., at what period did this symptom occur? Was it before taking the medicine he had hitherto been using? While taking the medicine? Or only some days after leaving off the medicine? What kind of pain, what sensation exactly, was it that occurred on this spot? Where was the precise spot? Did the pain occur in fits and by itself, at various times? Or was it continued, without intermission? How long did it last? At what time of the day or night, and in what position of the body was it worst, or ceased entirely? What was the exact nature of this or that event or circumstance mentioned - described in plain words? Hahnemann (1997), Organon, Paragraph 86

Paragraph 94 continues:

> While inquiring into the state of chronic disease, the particular circumstances of the patient with regard to his ordinary occupations, his usual mode of living and diet, his domestic situation, and so forth, must be well considered and scrutinized, to ascertain what there is in them that may tend to produce or to maintain disease, in order that by their removal the recovery may by prompted. Hahnemann (1997), Organon, Paragraph 94

I also find paragraph 98 remarkable:

> Now, as certainly as we should listen particularly to the patient's description of his sufferings and sensations, and attach credence especially to his own expressions wherewith he endeavors to make us understand his ailments - because in the mouths of his friends and attendants they are usually altered and erroneously stated, - so certainly, on the other hand, in all diseases, but especially in the chronic ones, the investigation of the true, complete picture and its peculiarities demands especial circumspection, tact, knowledge of human nature, caution in conducting the inquiry and patience in an eminent degree. Hahnemann (1997), Organon, Paragraph 98

About anamnesis, there are a several other paragraphs, and I find it astonishing *how exactly* Hahnemann instructed his students to record the patient's medical history.

Of course, our normal medicine also tries to build up a comprehensive medical history, but, quite honestly, it generally falls far short of this objective in practice. Doubtless the good old family doctor still exists, who is up against the economisation of his surgery. But in my experience, he remains an isolated case in today's medicine. And unfortunately, because of the strict organisation imposed by modern medicine, it is difficult for him to take the time really needed to discuss all the aspects mentioned above with the patient. In homeopathy, however, the essential thing is to dedicate oneself to the patient in precisely this detailed way - whether the medical history is targeted in breadth or in depth. It's about the person sitting in front of you, with all their peculiarities and details of the way they live, not just about some salient symptom. During the treatment, physical, mental, and psychological aspects are all taken seriously.

Different Aspects of a Holistic Homeopathic Anamnesis

- Physical
- Emotional
- Mental, Spiritual
- Social
- Family history
- Medical history

The therapist is looking for a way to identify a red thread or a pattern in all these factors which leads to a specialization of the patient's image and (in the traditionally homeopathic view), according to the principle of similarity, indicates to the drug to be selected.

In my experience, a model referring to the "levels of a disease" proved to be very helpful in ensuring that the anamnesis does not get lost in details and can proceed in a target-oriented way. I will introduce this model below.

4.1.6.1   Criticism of This Aspect of Homeopathy

Apart from the vagueness of the concept of holism, this is the least criticized of all the points mentioned, since it basically corresponds to the ideal of normal medicine (even though it can never be realized there in practice).

4.2   The Homeopathic Clinical Picture

When we talk about illness or symptoms in normal medicine, we almost always refer to a physical malfunction. Homeopathy, on the other hand, recognizes symptoms as an expression of a deeper disorder (of the vital force). Hahnemann sees the symptoms as an *"outwardly reflected picture of the internal essence of the disease, that is, of the affection of the vital force"* (Hahnemann 1997, Organon, Paragraph 7).

> There is, in the interior of man, nothing morbid that is curable and no invisible morbid alteration that is curable which does not make itself known to the accurately observing physicians by means of morbid signs and symptoms …
> Hahnemann (1997), Organon, Paragraph 14

I have talked about the difficult concept of the vital force before. Now I would like to deal with a question raised by Hahnemann's theory: If the symptoms are not the disease, then what does it mean to be sick? Or rather, how can one describe and experience illness? These questions have not only been pursued by homeopaths. There are also approaches to this in scientific medicine. In particular, the field of psychosomatic medicine is dedicated to the connection between physical and psychological problems. Depending on the school of thought, the different forms of psychotherapy see the causes of physical symptoms as being more or less rooted in psychological problems or former individualized imprinting.

In homeopathy, we have always tried to see external physical aspects (symptoms) as a kind of signpost that points us to the patient's deeper problems. In addition, we endeavour to include and treat inner mental and spiritual factors (Hahnemann 1997, 2016; Sankaran 2003, 2005, 2009). Everything the patient reports is true and can be evaluated and included in the treatment. So, he may also talk about his inner self. But what is this "inner self" or "interior" of the patient? How can we be more precise about this?

I shall try to explain the "interior" by referring to the "level model of a disease" expounded by Rajan Sankaran. This is not an idea of classical homeopathy, i.e., homeopathy, which goes back directly to Hahnemann (Sankaran 2005). Rather, it is an extension of the basic principles presented by Hahnemann. This level model is not shared by all in homeopathy, and is rather new for some homeopaths. Nowhere else have I found such a clear concept, and none that has such practical advantages for therapy decisions. Homeopathy generally assumes that symptoms are only an external sign of an internal disagreement, but in a far less concrete way than in this model. Indeed, it makes it easier than ever before to explain hitherto vague terms such as "holistic", "individual", or "inner". I have deliberately simplified the original model and, for example, omitted an "energy" level, since it does not take us further in terms of scientifically justifiable homeopathy.

From a homeopathic point of view, disease affects the whole human being and can be described on the following levels from outside to inside:

- Physical level
- Emotional level
- Spiritual level

4.3   Levels of Disease

4.3.1   Physical Level

In scientific medicine, the physical level is the main level and is usually the only one on which a disease will be considered and treated. It is generally equated with the diagnosis. For example, we speak of "pneumonia" and we know that this is an inflammation of the lungs. With a medical diagnosis, a disease is (usually) clearly differentiated from others. A diagnosis is always a concrete guideline; that's why it is so important to make a correct diagnosis, i.e., to give the disease the right name. Wrong diagnoses lead to wrong treatments. The name of a disease always refers to the disease itself; it is not specific to the patient who has the disease. We take this for granted, but it makes a decisive difference compared with homeopathic thinking and the other levels.

To make a correct diagnosis it is important to know which symptoms, data, and facts demarcate this disease from others. In the case of pneumonia, as doctors, we see typical symptoms: cough, fever, pain or shortness of breath, weakness, and possibly sputum. The facts and findings are as follows: bacterial or viral inflammation of a lung area; the X-ray may show corresponding changes. The laboratory values will show correspondingly higher inflammation parameters, and specific noises may be observed during auscultation (listening). Any typical case of pneumonia is accompanied by such symptoms, facts, and findings. These symptoms and facts must be differentiated from those of other lung diseases (e.g., acute or chronic bronchitis, asthma). The symptoms of a disease are specific to a diagnosis, not to a patient. This means that, in order to make this diagnosis, patients must exhibit the specific symptoms, data, facts, and findings of pneumonia.

In contrast to scientific medicine, homeopathy does not regard this physical level as standing on its own.[2] Rather, it is considered as an expression of the "disturbance of the vital force" that appears on a different, deeper level. The physical level is imagined in homeopathic thinking as a kind of canvas on which there appears a projection as in a shadow play. Yet the real thing happens elsewhere. Accordingly, a homeopathic anamnesis is more like going

---

[2]In the homeopathic sense, the so-called modalities also belong to the physical level. Modalities are defined as criteria that influence or are related to physical complaints, for example: Which influences cause improvement or worsening? Is there a characteristic (daily) course of symptoms or other anomalies? Modalities are usually somewhat more individualised than the mere facts, sometimes also specific to a certain homeopathic patient and drug image. In scientific medicine, modalities don't play such an important role.

on a search with the patient to find the underlying cause that has led to his or her symptoms. Although symptoms are recorded in homeopathy, they are only relevant in connection with the patient's inner state and condition. Along with this inner state, they form the patient picture.

The physical level is the level of a disease on which we can collect measurable and quantifiable findings in the form of figures and data. Blood pressure, body temperature, blood values, ECG derivations, thyroid gland size, etc., are objectifiable values. This distinguishes the physical level from the other two levels, which are far less tangible.

4.3.2 Emotional Level

The physical level is followed by the level of emotions. Homeopathy assumes that diseases can trigger emotions, and conversely that they can also arise from emotions, or are related to them in other ways.

First, I would like to define the term emotion. For this purpose, here is a short definition, translated from the German Wikipedia, which will be useful in the context of our reflections:

> Emotion (from Latin ex = out and motio = movement, excitation) is a psychophysiological or psychological process which is triggered by the conscious and/or unconscious perception of an object or situation and is accompanied by physiological changes, specific cognitions, subjective emotional experience, and a change in the behavioural willingness.

Emotion is therefore a (conscious or unconscious) reaction to a stimulus with an "outward movement". An emotion is not just a feeling, but a psychophysiological process. Emotion is coupled with physiology (the physical level). It has an influence on the physical level and can in turn be influenced by physical processes.

Emotional complaints are already somewhat individual and no longer have a specific connection with the disease. While the diagnosis of pneumonia with its specific symptoms may trigger anxiety in one patient, for example, another may be relieved and glad to lie down in bed for a few days. Another patient may be convinced that the current stress of his divorce has led to the disease. (Such a subjective assessment can be accepted as a legitimate feature in homeopathy and be taken as is.)

On the emotional level, patients express their inner stress individually from a homeopathic point of view. Feelings, fears, worries, hopes, desires, and their personal history play a role in the homeopathic anamnesis (always in connection with the symptom). Traditionally, in the repertories emotions were represented under "states of mind" or "symptoms of mind". The emotional state or conspicuous peculiarities in this area are part of every patient image. The homeopath explores this individual emotional level by asking openly ("How do you feel about symptom X?"), through active, empathic listening and without judging what is said. The time therapist and patient spend together, and the deep doctor-patient relationship are important. The aim is to increase trust between the interlocutors. It is not only a matter of systematically requesting findings and sensitivities, but also of exploring the feelings that the patient associates with them. If, for example, a patient says that he or she realizes that the symptoms always occur under stress, we try to find out how the patient feels under stress, what stress means to him or her, what emotions he or she associates with it and with the symptoms, and what individual psychophysiological connection there may be with the physical symptoms.

Psychotherapeutic methods work, among other things, on the individual emotional problems of patients. Psychosomatics are responsible when the interaction of emotions and physical symptoms is concerned. Here, for example, the issue is to reappraise the individual's past, specific traumas, etc. An attempt is made to work out an individualized imprinting, to make the patient aware of it through conversation, and to make it accessible to change in this way. This approach is already more specifically tailored to the patient than to a disease. Two patients diagnosed with phobia (anxiety disorder) will surely have different in biographies and emotional conditions that require individual psychotherapeutic (and, if necessary, individual medicinal) treatment. These psychotherapeutic forms of treatment are closer to homeopathy than to purely scientific medicine, as they accept and treat the emotional level and assume a multi-causal pathogenesis. Psychotherapy, however, tends to focus on the "Why?" or "Where does it come from?" while homeopathy focuses on the "What exactly is your experience of it?" or "What is the connection between the symptoms and precisely this patient, and what do they mean for the patient picture?

By the way, research in the field of conversation in doctor-patient interactions comes to a very similar conclusion to the one suggested by Hahnemann:

Every patient is an informant, but not every informant is a good informant. Whether he is good or not depends on the examining physician and how he knows how to open up and experience the informant's horizon of thought and experience. (Kohnen 2007)

Contribute to the patient's improvement by giving care according to the biopsychosocial model, by practicing a narrative interview style, by letting the patient tell the story, by (…) actively listening to him/her, if possible uninterrupted, and repeating literally or paraphrasing his/her words at each suitable opportunity. (…) The art of conducting medical conversations lies in the doctor's verbal interventions achieving a communicatively accurate fit, (…) which should not be arbitrary but context-sensitive. (Koerfer et al. 2008)

These points make it clear that lasting changes in the quality of communication, which could establish the dialogue between doctor and patient as the "heart of medicine", can only be achieved in the long term and through concerted action by healthcare policy and a paradigm shift in university medicine. (Nowak 2010)

However, by calling for such a style of conversation, homeopathy already now tries to fathom the emotional level of the patient.

4.3.3 Spiritual Level

If you should stumble over the term "spiritual" when you first see it, I would ask you to read the section below entitled 'The terms "spirit" and "spiritual" ' in order to understand what I mean by this and why I chose this term.

- "I'm in so much pain while coughing, as if someone's trying to ram a knife into my ribs."
- "I'm afraid I'll be coughing up lungs any moment now, the cough is so bad."
- "I can't breathe, it's as though there's a heavy weight like a rock on my chest."

Have you ever heard descriptions like these from conversations with others, or given them yourself? In our everyday language we often use images and metaphors to describe and explain diseases or symptoms. In normal medicine, such metaphors are meaningless. Scientific physicians shake their heads, usually compassionately, when patients describe their symptoms so graphically. It is not (yet) clear to us how this level might be valuable. Here the description becomes highly individual and therefore highly "distinctive" or "peculiar" in Hahnemann's homeopathic sense:

> [...] the more striking, singular, uncommon and peculiar (characteristic) signs and symptoms of the case of disease are chiefly and most solely to be kept in view. Hahnemann (1997), Organon, Paragraph 153

What can we conclude from this? Two people with the same diagnosis of pneumonia, with similar symptoms and emotions, may not at the emotional level use the same image to describe their feeling of illness. Scientific physicians will inevitably switch off at this point, but for the homeopath it's just here that things begin to get interesting. It is fascinating to note that patients seem to realize very clearly how they themselves imagine their disease on a metaphorical level - and they really experience it that way. Every day in my practice I experienced the confidence with which patients move on this level. Especially with the help of the sensation method, a specific, highly individualized sensation can be established in connection with the symptom.

What is "sensation"? This is not a specifically homeopathic term. Again, I would like to use a good explanation from the German Wikipedia:

> Sensation is understood as a precondition of perception and a first stage in the neuronal processes that ultimately enable perception (through the senses). (...) In contrast to apperception (= clear and conscious perception), sensations can therefore be processed subliminally (unconsciously) and vegetatively, too.

> Sensation is the successful result of neuronal excitation and "first" becomes effective (so to speak) in the primary sensory areas of the brain, before entering consciousness as a specific perception in other - secondary and tertiary - areas of the brain.

Sensation is thus a process that precedes perception, often taking place and being processed unconsciously. If I have the feeling that an imminent situation is very threatening (e.g., a job interview), I may actually experience it this way. This is obvious and has no medical value in itself. But let's go one step further and ask our patients quoted above the following questions:

- How can I imagine exactly what you're talking about, or what you mean when you say "it's like someone's going to ram a knife into your ribs"?
- You're afraid that you're about to cough up your lungs. Tell me more about that. What exactly is your experience?

- How exactly do you experience the feeling of a stone lying on your chest? How do you perceive that? What sensation does it induce in you?

A sensation is not a feeling or an emotion, and therefore - according to homeopathy - it is a further level in human consciousness. A sensation expands the feeling by the dimension: How do I perceive this feeling in me? How do I experience myself when I have a feeling (e.g., fear)? Such a sensation can be expressed in words and is also recorded in the homeopathic anamnesis. In this very individual sensation, we find - from the perspective of a homeopath using the sensation method - the spiritual core of a disease when this sensation can be generalized and broadened into a sensation recurring in all areas. Only then does the concept of sensation become relevant to homeopathy.

If the patient perceives a job interview as threatening, and in addition this sense of threat is a fundamental issue in his or her life, then we will have identified a homeopathically interesting vital or core sensation (Sankaran 2005).

The Core Sensation

Basically, the aim is to identify not only a situational sensation, but a *general* sensation for the patient. From the spiritual level of sensation on, this is all about a principle that is generally valid in us, i.e., the sensation is not limited to the particular illness.

Homeopaths postulate that this core sensation is upstream to our entire perception and that we perceive our whole life and the outside world through this - including also the illness. This may allow us to recognize an entire "false perception" via the disease, as though the disease were a kind of keyhole.

Homeopathy assumes that such ideas create an unconscious state of stress. If someone feels permanently threatened inside, he or she will never experience this as relaxing. In this respect, such unconscious perceptions and sensations - seen from a homeopathic point of view - may have an influence on the stress levels of a patient and thus lead to health impairments in the long term. Conversely, such sensations and perceptions can influence diseases. More about this later.

This mental level is considered highly specific *to the patient*, but not to the particular disease.

The most specific feature of homeopathic conversation does indeed concern the spiritual level. Here, hitherto unconscious sensations, ideas, or beliefs can already be worked out during the conversation. Where appropriate, a generalizable core sensation may also be discovered. In homeopathy, such core sensations are considered as the cause of a patient's increased stress sensation, because they do not allow him or her to perceive the outside world and himself in a neutral manner.

In the best case, reaching an awareness of the discrepancy between reality and perception leads to self-knowledge; the method assumes precisely this. That sensations can have an influence on the body is also accepted in psychological circles.

Karl Jaspers describes certain anomalies of object consciousness which can be understood as changes in perceptual ability, or a "disturbance of sensation". Regarding such "perceptual anomalies", he differentiates between changes in intensity and quality on the one hand and perceptual abnormalities on the other. These disorders can also manifest themselves in anomalies of feeling and states of mind or as *disturbances of basic psychosomatic facts*.

However, such knowledge has not yet been made part of medicine, at least not for the purposes of therapeutic use. In homeopathy, on the other hand, the main focus is on the individual sensation of the patient. With the help of the therapist, it is hoped that a situation-appropriate change of self will take place as result of this self-knowledge of the sensation.

Traditionally, these spiritual statements were viewed under the headings of "delusional ideas" or "as-if symptoms". In this context, the term "delusional idea" was not understood in the psychiatric sense, but rather as a term for particularly personal sensations. The term "as-if symptoms" is used because many patients use phrases like "It's as if…" (for example, " … as if I were trapped").

Scientific medicine has no equivalent for this level. At best, it knows patients whose perception has shifted so morbidly that they can no longer dissociate themselves from it (e.g., paranoia, phobias, schizophrenia). Other spiritual teachings, healing methods, or ideas also describe this kind of individual "upstream" sensation, which can influence our entire perception, but they are not part of medicine. Many such spiritual teachings aim to free us from such individual sensations (designated as "sets of beliefs"). In the ideal case, homeopathy provides us with a very concrete approach to this level. However, this level has not yet become part of medicine. As a normal doctor, I wouldn't be able to take such individual feelings and statements (symptom-related or general) into consideration. With the sensation method, on the other hand, it is possible to detect and track down a generalized sensation that was hitherto upstream of the perception and - according to the homeopathic doctrine - has led to stress. The question of whether the sensation method is similar or even superior to other psychological methods (e.g., acceptance and commitment therapy – ACT[3]) goes beyond the scope of this book.

An Aside for Homeopaths Who Are Familiar with the Sensation Method

Unlike Sankaran, I do not distinguish between *delusion* (imagination, delusional idea) and *sensation*. I don't go along with the idea that sensations in human beings are caused by the "energetic" connection with a drug. That's the part of homeopathy I would like to leave behind. I acknowledge that the sensation method provides access to specific human sensations and ideas which the patient may not yet be aware of and which may stress him/her. I consider sensations as an extension of human imagination. In this respect, for example, a patient may feel "trapped and constricted" and describe this sensation very precisely and in detail. However, it remains a *sensation*, created in the patient's imagination and it is by no means equivalent to the *"spirit of the drug"*. I don't share the view that the patient speaks at the sensory level *"as his drug"*, directly from the source, so to speak. I therefore don't share the classic view that the patient is talking about a non-human sensation, namely that of his medicine.

From a therapeutic point of view, we can make the patient aware of his or her sensation, and accompany and guide him or her in this self-knowledge.

However, we cannot claim to know how an Anacardiaceae (poison ivy) or an eagle (even less so electricity, dog's milk, or tubercle bacteria) experience a "sensation" and how they speak to us as a "source". These are human attributions (anthropomorphisms) which we are making here and which we can perhaps *imagine*, but which are certainly not facts.

In my opinion, we can't go beyond the delusional level without the need to provide some kind of explanation.

4.3.4 Levels of Disease – Summary

In Table 4.1 I summarize the levels of a disease and their peculiarities.

We work through the levels from the outside to the inside, as it were. In homeopathy, the way this happened and happens is not always as clearly differentiated as shown here. Not all homeopaths use this way of thinking about it. Rather, a colourful mixture of the three levels often flows into the patient picture, and the final assessments are not always free of interpretations on the homeopath's side. This is often simplified or generalized, and often

---

[3]https://en.wikipedia.org/wiki/Acceptance_and_commitment_therapy.

Table 4.1 Levels of a disease

Physical level	Less individual, specific to the disease	Main level in scientific medicine
Emotional level	Relatively individual	Main level in psychotherapy. Psychosomatic medicine focuses on the connection between this and the physical level
Spiritual level	Increasingly individual, specific to the patient	Main level in homeopathy and spirituality, partly also in psychotherapy

influenced by the need to find the symptoms in a repertory, which can lead to the same kind of stereotyped thinking as in scientific medicine. Anyone whose symptoms get worse through warmth and washing, who is often annoying and opinionated, and who suffers from the delusion that he/she is more often right than others is simply a sulphur patient. But anyone who is sensitive and weepy, who likes to be loved and cuddled, who likes ice cream but poorly tolerates it and isn't thirsty is a pulsatilla type. I don't consider this way of thinking as sufficiently individual and open-minded. Homeopathy has no advantage over other methods here, even though levels other than the purely physical one are involved.

I am aware that in scientific medicine there is no attempt to include these other levels. Disease is primarily a physical problem in that case. But I find it fascinating that homeopathy attempts to include further aspects. Certainly, this is still to some extent rather unsystematic and borders on the esoteric. Nevertheless, I still find it interesting, especially since it seems to be the most important feature that patients hope to find among homeopaths. For myself as a therapist, the benefit of this division of levels is the clear distinction between spiritual-mental problems and physical or emotional ones. The deeper the level we reach in a patient interview, the more individually the problem presents itself to us and the more holistically we understand the patient's problems. The patient will feel that she/he is being taken seriously in her/his whole being and that her/his core problem is indeed being treated "holistically".

Benefits of the Level Model

If a disease is divided into the levels at which a patient's main problem occurs, it can be described in more detail and individually than was previously possible in medicine. The new feature is that this also offers access to the spiritual part of the human being.

This approach makes it possible to move forward from a disease-specific diagnosis to a patient-specific diagnosis ("core sensation").

While scientific medicine always focuses on symptoms, homeopathy practised in such a holistic way focuses on the patient and on his or her individual core problem. At the same time, however, thinking in levels would also mean treating diseases with a very clear material cause (e.g., bacterial cystitis, proven microbiological findings) physically (e.g., administration of an antibiotic).

However, Hahnemann writes that it is "only the vital force that is disturbed by diseases. How can this be reconciled with the "levels of a disease?

> It is the morbidly affected vital energy *(i.e., vital force)* alone that produces disease, so that the morbid phenomena perceptible to our senses express at the same time all the internal change, that is to say, the whole morbid derangement of the internal dynamis; in a word, they reveal the whole disease … Hahnemann (1997), Organon, Paragraph 12

Where do we encounter this vital force in the level model?

On the spiritual level, diseases can also be experienced as spiritual figments of imagination that are highly individual and peculiar. It seems clear that a vital force can be imagined here as a mental idea.

If "the vital principle … derives all sensation" (Organon, Paragraph 10), then the disturbance of the vital force can be found via this sensation. Spiritual problems are problems of imagination or perception which we are usually not aware of. Hahnemann has called these problems, rather vaguely, "disturbance of the vital force" (nowadays we refer to "stress", an equally vague term). I accept this as a spiritual figment of imagination, not as a fact or even a physical quantity. Then it is a metaphor on the spiritual level - where we can easily accept it. And, not unimportantly, we can take this sensation into account in a treatment by contrasting it with new ideas and information (e.g., "The globules will help me to regain my health and guide the path of healing to my goal"). I will discuss this in detail later.

4.4 The Terms "Spirit" and "Spiritual"

In Hahnemann's texts in the Organon, the problematic words "spirit", "spiritual", and "spirit-like" are often encountered (in Boericke's translation of the Organon, these are also referred to as "dynamis" and "dynamic"). I will use these terms in this book. Of course, we can immediately associate nearly everything we can think of with such terms. By this, homeopaths probably mean something superordinate, something qualitatively superior to the purely material. Critics, however, cry out that this is too reminiscent of spiritual healing or even necromancy - i.e., it's the kind of humbug we don't want to deal with in medicine. For example, I recently used the word "spirit" on an internet discussion platform on homeopathy and was surprised when a critic replied: "There are no ghosts". To avoid such misunderstandings, I would therefore like to define the word "spirit" precisely. Again, I would like to refer to a summary of the term "spirit" (and not "ghosts") in the German Wikipedia:

> The cognitive abilities of humans are usually referred to as "spiritual". This includes perception and learning as well as remembering and imagining, fantasizing, and all forms of thinking (e.g., selecting, deciding, intending and planning, predicting, assessing, evaluating, and observing). This also includes the consequently necessary vigilance and mindfulness as well as concentration of all kinds. These range from hypnotic and other trance-like states, on the one hand, to those of monitoring and maximum presence of mind, on the other.

Thomas Metzinger (born 1958), a German philosopher who seeks an exchange between philosophy, neuroscience, and cognitive science and who has devoted much attention to the philosophical interpretation of the search for neuronal correlates of consciousness and mind, writes:

> Our scientific-philosophical self-image (is) in a state of fundamental upheaval. Our theories about ourselves are changing, especially the image of our own mind (…): genetics, cognitive neuroscience, evolutionary psychology, and the modern philosophy of the mind are gradually providing us with a new image of ourselves, an ever more precise theoretical understanding of the (…) deep structure of our spiritual dimension, its neuronal basis, and its biological history. We are now beginning - whether we like it or not - to understand our mental abilities as natural characteristics of ourselves, as qualities with a biological history that can be explained by the methods of science (…). (Metzinger 2013)

I use "spirit" only in this sense and consciously distance myself from any esoteric or religious interpretation, or indeed the originally rather nebulous homeopathic interpretation:

> The spirit is the creative instance of consciousness in a person, which is superior to the mind and at the same time explained by scientific principles.

We may thus say that, although the human spirit cannot simply be reduced to physics and biology, it is still in harmony with natural science.

The scientific principle of emergence states that the whole can be more than the sum of its parts, i.e. that the appearance of new, more complex properties is possible through interaction of lower order factors. Many cells give rise to a human being, but such a being can no longer be understood as a mere cluster of cells. From physics and biology (and physiology) human consciousness emerges, along with thoughts, feelings, and the spirit as well. Can the big picture be derived solely from the individual parts, or reduced to them in some way? This is a long-standing philosophical problem with a whole spectrum of possible answers, from "Absolutely" to "No, of course not".

In the natural sciences and humanities and their treatment of the human being as a subject of investigation, the prevailing opinion is that there can be nothing unnatural. The phenomenon of the spirit isn't deducible one-to-one from physics, but it doesn't contradict it either. The "surplus value" of the larger whole is derived from the synergies (interactions) of the basic individual entities, which, however, will preserve their own logical consistency. This is not easy to understand; the science of evolution has also been trying for a long time to find consistent hypotheses to explain this. (Schmidt-Salomon 2014 - Wikipedia, keyword: Emergence). What is clear is that, over the course of many millions of years, intelligent and sentient human brains have developed from rather simple single-cell organisms in a kind of primordial soup. But this development didn't at some point suddenly allow the human mind to appear by some kind of miracle. In fact, complex changes have taken place over the eons, but always respecting the laws of evolution, physics, and biology. The complexity and intelligence of human brains today allow us to explore ourselves, and there are indeed many aspects of the human being that we can today explain in an absolutely satisfactory way. In contrast to Hahnemann's time, we are no longer obliged to accept things like a vital force, spiritual energy, etc. This is how I personally perceive the human mind, and I would like to see it receive more attention and more consideration in medicine. Homeopathy offers a small step towards this, because it attempts to include the human spirit - but no longer in the "spiritual" sense today.

Our human spirit enables us to develop ideas and perceptions or to be influenced by them.[4] The spirit enables us to grasp complex interrelations

---

[4]This is one of many possible positions in a wealth of (para-)scientific, psychological, and philosophical considerations. I have decided in favour of this position for pragmatic reasons, because the explanations discussed later as to why homeopathy may have real effects can be built upon it.

and to develop ideas. For example, when we solve a mathematical task, it is an achievement of the mind, i.e., an achievement of thought. However, if we use this mathematical task to estimate the statics of a building and make it predictable, this is a complex mental achievement. The building doesn't exist yet; it's in the planning stage, only in the planner's head. It's an idea. Until the completion and examination of the statics, it remains an ideal concept that can be changed by planning, thinking through, deciding, and assessing in the brain (in the spirit). Without the spirit, you could read this book, but you couldn't gain any personal benefit from it, or form an opinion about it. The spirit is more than the mind or mere thoughts; the creative part of it makes it superior to the mind. In the human spirit we can create and understand ideas, images, and metaphors.

The spirit is extremely important in homeopathy, as described more precisely in the discussion of the "levels of a disease". I would like to just emphasize here that the human spirit can be affected by ideas like "This will help me". With homeopathy it may be possible to use this influence as a therapy - if we consider the spirit as a normal part of our human existence, alongside the physical and emotional part. We should therefore free ourselves from the concern that the term entails something esoteric or vague. The idea that health also includes spiritual aspects, that the spirit of the person is involved in the case of an illness, is (still) an unfamiliar idea in normal medicine. I shall say more about this below.

4.5   Homeopathic Medicines and the Placebo Effect

Homeopathy is traditionally a drug therapy and it also used as such by patients. They are usually looking for natural, gentle drugs with no side effects in the hope of being able to do something for themselves and their health outside of normal medicine. (I have already discussed the problem of this understandable desire in the sections entitled "Being able to do something" and "No side effects".)

Traditionally, patients in homeopathy are prescribed drugs that are selected according to the principle of similarity. In homeopathy, "similarity" means that a medicine that causes certain symptoms in a healthy person can cure a disease. This is done by causing a similar "artificial disease" in the patient, which his or her body processes as a first step in overcoming the actual disease. Homeopathic drug pictures give a detailed description of the condition brought about in healthy people by taking the given remedies and the indications for which it can therefore be prescribed. This concerns physical as well as emotional and spiritual aspects. But how can we benefit from this idea? Can this basic homeopathic principle still be maintained even if we have found that homeopathic drugs do not contain anything "similar" at all, but in fact contain nothing, and that as a result the drug images produced by means of the drug trials have been generated by a placebo and are therefore highly questionable?

Hahnemann's theory that in fact a "drug energy" or "drug information", similar to the "disease energy", becomes effective via the globules cannot be maintained. In this sense, the hypothesis of similarity must be completely discarded.

However, other aspects could be helpful in a figurative sense:

- First, it is helpful for the patient to make the following observation: "There is help for my complaints. I'm not just imagining it. The homeopath recognizes a similar pattern, familiar to him, that can be treated." This certainty alone can be helpful. I suspect that there is even more going on at the spiritual level. At this point, it may be particularly helpful that there is "nothing real in the globules". This may allow room for exactly the information that the patient needs. I'll come back to that.

- The homeopathic therapist is open to the idea of identifying a specific spiritual problem or to guiding and accompanying the patient in detecting such a problem, starting from the main symptom. The homeopath is sure that there is a similar picture or drug for it, and this assurance is conveyed to the patient. Since the homeopath knows that the spiritual level of a disease really exists, and does not shy away from it, he/she can also recognize it, and hence treat and accompany the patient at this level.

- Patients who no longer need to feel that they are "crazy" if they turn to this "core" aspect of their complaint will be free to observe themselves more closely, more benevolently, and more comprehensively, and hence perceive themselves in a better light. They can thus identify their needs more clearly and act accordingly to improve their situation.

These points could be examined in more detail; they have nothing to do with the original principle of similarity between drug image and patient image. What the patient can take home from the homeopathic anamnesis, however, is a consistent overall picture in the sense: "Oh, that's what it all has to do with me".

The Special Relationship Between Homeopathy and the Placebo Effect
In a homeopathic conversation, we work out on which level the patient feels his or her core problem and then seek to treat exactly this with the globules. The globules therefore always carry exactly the required information. Because it is our message as homeopaths for the patient: whatever we are told, it makes sense for us and can always be translated into a very individualised and specific medicine. If the focus is on physical symptoms, we prescribe globules, for example, against the severe, paralyzing back pain that is particularly intense at the beginning of a movement. On the emotional level, we prescribe the same globules for the concern associated with it, and also on the spiritual level, for the sensation of being generally trapped and paralyzed, and wanting to be free and agile instead. To change or become conscious on the spiritual level of a disease, which is the level of images and imagination, images and imagination are helpful. Imagining that "This will help me" - will help. This is now exactly the definition of the placebo effect.

Here is a short definition of the notion of "placebo" from the German Wikipedia:

A placebo (Latin for "I shall please") is something presented as a medicine but which does not contain any active ingredient and therefore cannot have a pharmacological effect caused by such an ingredient. Placebo effects are positive changes in the subjective state of health and in objectively measurable physical functions attributed to the symbolic significance of such a treatment. They can occur for any type of treatment, not just in sham treatments.

Looking at the definition of the placebo effect, there can hardly be any doubt (since we have made it clear that homeopathic medicines are materially ineffective) that they act through this effect. They contain nothing but the message: "I'll help you with your complaints."

The placebo effect does not mean that a drug is ineffective - only that it does not work through pharmacological ingredients.

There is nothing in it, and yet something happens that we can see and feel as patients. In this respect, one can agree with homeopaths when they say that homeopathy works! And one can imagine that it should work all the better, the closer the "drug" comes to the exact individual subjective perception of the core problem. The effect should be greater, the more the patient has the sentiment: "It will please *me* - it will be good for *me*". Patients now receive *their* globules and believe in their efficacy, *which is precisely tailored to them and their specific complaints* - and, lo and behold, they work. This effect is further enhanced by the therapist's diligent efforts to choose the right medication.

If a drug doesn't work, another one is selected; the patient may have lost a bit of confidence, but most of the time hope prevails, and the placebo effect once again brings about a change. Apparently, this placebo effect can also be repeated, when patients take their medication at home (acute illness or acute relapse) as part of self-treatment. The more precise and the more complicated the prescribed dosage, the greater the placebo effect (as we know from conventional placebo studies). Here is an example:

> Stir three globules (clockwise!) in a half-full glass of tap water until they dissolve and take with a wooden or plastic spoon one teaspoonful of this solution every two hours after stirring vigorously again. This should be alternated with two globules placed directly on the tongue until they have dissolved; every 4 h, every other day. Avoid coffee, mint, and other strong influences during this time, and make sure you get enough sleep.

One of the main arguments against a mere placebo effect in homeopathy is that the globules also work for children and animals. It is said that children and animals would not benefit from the homeopathic setting or the positive placebo idea ("This will help me") and therefore cannot be influenced by a placebo effect. However, this is a fallacy. Just the normal attention by the caregivers (father or mother, pet owner) is sufficient to trigger placebo effects. In addition, the caregiver's personal mood, the relief of being able to do something for those in need of protection, and the expectation of an improvement will be transferred to the child or animal. This effect is called "placebo by proxy", i.e., the confidence conveyed by the therapist or veterinarian to parents or the pet owner is transmitted indirectly to the child or animal and thus triggers the effect. Children's and animals' great sensitivity to the moods and sensitivities of caregivers works to enhance the effect. Basically, attentive parents and caring pet owners will confirm this as an everyday experience. The alleged impossibility of a placebo effect in children and animals therefore fails as an argument for a specific effect of homeopathy.

Let me make things clear once again: homeopathy would like to be part of medicine so it must prove its effectiveness in scientific studies, since this is the common procedure. Individual observations are gratifying, but no proof of an effect. No study has ever found an impact beyond a placebo effect, and the renowned medical journal "The Lancet" already carried the title "The end of homeopathy" in 2005 - after the findings of a major study (Shang et al. 2005; The Lancet 2005). But this has not yet changed anything in the way homeopathy is perceived or carried out.

Let's return to the "information", which, according to homeopaths, is supposed to be contained in the globules. The globules do indeed contain information. However, it is not a mysterious piece of "information", but rather an *actual and literal* piece of information. They always contain exactly the information (or rather the meaning) that the patient needs. If an illness can be described and experienced as a spiritual problem at level three, then it can in principle be changed by some form of spiritual imagination, i.e., an idea or information on the same level. There is therefore no need for drugs to cure it.

> The globules are carriers of meaning and individualised autosuggestion, not drugs in the proper sense.

Only in this sense can we still accept Hahnemann's idea today.

However, to identify this spiritual part of the disease, the sensation method in homeopathy or other modern methods are needed. An investigation into symptoms alone, filling out a questionnaire, or self-treatment using quick-finders will not be sufficient. As homeopathic therapists, we lead the patient to a state of spiritual self-awareness starting from the physical symptom, going through the emotional level, and then ideally providing a way for the patient to undertake self-change. This would be homeopathic therapy without homeopathic medication and would (presumably) go beyond a placebo effect. It remains open, however, whether other psychological methods might not be far superior to homeopathy in this and whether they might not already present valid results.

Further investigations along these lines would require a prior consensus between homeopaths, because as mentioned before there are so many differences within homeopathy, and so many ways to carry out the anamnesis.

In my daily practice, the result was that I would increasingly provide my patients with an exactly named core mental topic which I had worked out previously by the sensation method. In my introductory example, this meant that I would discuss in detail with Mrs. M. the idea that her basic problem on the spiritual level consists of a "general captive and restricted feeling". For this problem I would give her a homeopathic medicine in the form of globules and "with" them the literal information about her basic problem. She would attribute the change in her life to the globules. In my opinion, however, that change would be due to the placebo effect and our therapeutic setting.

Perhaps some patients are just as happy with an improvement due to a placebo effect, rather than the pharmacological impact of a conventional medicine. But in this respect, too, we must at least be honest. A placebo effect does indeed imply that real physiological changes can take place. But homeopathy cannot achieve more than that. None of the studies conducted so far has come to any different conclusion. We homeopaths should not try to convey anything other idea to the patient. Whether the globules, if taken with a very individual message as autosuggestion carriers, could develop a kind of super-placebo effect remains to be seen - through research.

4.6 What Can Homeopathy Do That Medicine Cannot?

Once again, I summarize the advantages and peculiarities of homeopathy (from the point of view of patients and homeopaths). I am not yet concerned here with the scientific examination of the assertions made. This will be done later. The effects described can occur in a well-managed homeopathic conversation. The diversity of homeopathy, however, makes a uniform assessment difficult, so at the present time, we cannot speak for homeopathy in general.

In 1921, in his preface to the Organon (*apparently not published in English*), *Richard Haehl*[5] already noted the idea that there was no need to take Hahnemann's idea of homeopathy literally in all its aspects, but that it might nevertheless make sense when interpreted in some other way:

> Hahnemann undoubtedly had a highly developed intuition, and some of his statements, which a few decades ago appeared strange, perhaps even absurd, now seem quite reasonable and rational to the medically or biologically thinking physician. Even the "dynamis", the "derangement of the vital force", whereby Hahnemann simply wanted to express that for him an illness was not merely a disease involving some particular organ, but the participation of the entire organism as result of a disturbance of the body and soul of the whole person, has unexpectedly found confirmation during the last few years (…).
> *Haehl*, Organon, editor's foreword (1921)

---

[5]Richard M. Haehl 1873–1932 MD, a German orthodox physician who converted to homeopathy, studied at the Hahnemann College of Philadelphia, and wrote a biography of Samuel Hahnemann.

Unfortunately, this interpretation of Hahnemann's homeopathy has received very little attention so far. I myself only discovered this text in the Organon when working on this book, although I should have noticed it many times before.

In this respect, homeopathy may offer an approach that scientific medicine does not: healing should take place on various levels, going far beyond the physical area. This could correspond to many patients' desire for a holistic, individual, and empathic approach.

Is it true, as Paracelsus[6] stated that he who heals is right? Or shouldn't we say rather that he who is right heals?

Scientific medicine can clearly set out its knowledge and principles; it learns from its mistakes and can boast a host of significant healing successes. Normal medicine is therefore clearly right in terms of science, research, medical facts, and pharmacological and physiological fundamentals. Its healing effects are proven by many studies and are unquestionable.

Homeopathy, on the other hand, may be right in its concept of the human being and its understanding of disease, which are complex and holistic, including individual feelings, thoughts, sensations, and even unconscious states. Here we have discussed the effect of the therapeutic setting, the possibility of increased self-awareness, and the success of the placebo effect (which can also be used as a kind of autosuggestion). However, homeopathy has not as yet been able to prove any of this.

Why do I emphasize the homeopathic image of illness and the human being? Allow me a small digression. The World Health Organisation's definition of health as the first of its principles is as follows: "Health is a state of complete physical, mental, and social well-being and not merely the absence of disease or infirmity." A state of complete well-being seems to me to be a very high standard and practically impossible to achieve. With scientific medicine alone, however, it would clearly not be possible to reach such a state, since it does not include the spiritual level and neglects the emotional level (and the social level even more so). Unfortunately, we will fall far short of the WHO principle if medicine largely reduces people to failures of physical function and if it treats them solely according to standardized clinical guidelines that do not provide for individualization. This could not do justice to the differences between patients.

Because of this, we need a concept that encompasses body, emotions, and spirit. Whether homeopathy is suitable for this remains to be seen. In any

---

[6]Paracelsus (1493/4–1541), born Theophrastus von Hohenheim, Swiss physician, alchemist, and astrologer, pioneered several aspects of the Renaissance "medical revolution".

case, to some extent it offers a system that would rather do justice to this claim and is well received by patients. Homeopathy makes it possible to find out at which level (physical, emotional, or spiritual) the patient's main problem is located. It is an approach that offers the opportunity to provide healing and support to the patient at the level where the homeopath sees the individual core problem. For some, the physical problems may be in the foreground, while for others it could be their spiritual health that requires attention. So far, scientific medicine has specialized primarily in the field of physical problems.

What Hahnemann described as a healthy state can certainly apply, even if it does not allow for *complete* well-being (although once again we stress that the "spiritual vital force" here is only to be understood as an *idea*):

> In the healthy condition of man, the spiritual vital force (autocracy), the dynamis that animates the material body (organism), rules with unbounded sway, and retains all the parts of the organism in admirable, harmonious, vital operation, as regards both sensations and functions, so that our indwelling, reason-gifted mind can freely employ this living, healthy instrument for the higher purpose of our existence. (Hahnemann 1997, Organon, Paragraph 9)

Homeopathy is often rightly accused of preventing the application of medical therapies which are said to be incompatible with homeopathy. Thanks to the level model of a disease, this accusation can be viewed in a more differentiated way. It is quite possible to prescribe an antibiotic for pneumonia at the physical level and to carry out homeopathic therapy at a higher level. Since the therapist knows at which level the patient needs or wants to be treated, the therapies do not contradict each other. On the contrary, they complement each other.[7]

Unfortunately, this is not what happens in practice; I am painfully aware of this. So far, scientific medicine has no concept for those parts of health that don't take place on a physical level (and sometimes little interest in it). Here, homeopathy as a method, if suitably reconsidered, can help us to extend our notion of medicine. It could also thereby provide care for those 75% of patients who make use of alternative medicine, often without the knowledge of their general practitioner or specialist. It is a striking fact that homeopathy, which had already become almost extinct, experienced a renaissance in the

---

[7]Even the coffee abstinence which is often demanded during homeopathic treatment (for reasons that are no longer comprehensible today) can be put to rest by the level model. For how could coffee, for example, interfere with the awareness of a core spiritual problem? The fact that it makes sense in general to pay attention to overuse of stimulants speaks for itself and for a healthy mind.

1990s. Shouldn't this give rise to reflection? It happened - in Germany - at a time when a flat-rate reimbursement system for doctors and clinics was increasingly being introduced, forcing many doctors and clinics to work under completely different conditions than before. The system was introduced at that time to counteract the skyrocketing cost of healthcare. This has indeed succeeded in some cases, but at what price? The length of hospital stays has been greatly reduced and the average treatment time per patient has been reduced to a minimum. Here it is not the doctors who should be accused. For the main part they only try to cover their costs. It is rather the system that should be questioned. However, it is not my intention to reproach anyone; I just want to draw attention to the fact that a problem may have been caused here.

Normal medicine has outsourced the patient; it treats "cases". Is it surprising that patients seek a more human treatment elsewhere? Even if this is practised unscientifically and sometimes even in an irresponsible way? Might we also conjecture that scientific medicine could drive patients into untrustworthy hands when practised in this way? And wouldn't it be a good idea to bring human aspects back into everyday medical life?

For example, who knows more about a patient than a homeopath after a complete homeopathic anamnesis? Recall that the homeopathic anamnesis includes:

- physical,
- emotional,
- mental,
- spiritual,
- social,
- family history, and
- medical history.

As part of this anamnesis, the homeopath will also work out on which level of a disease the patient wishes to be cured or accompanied. Recall that a disease has:

- a physical,
- an emotional, and
- a spiritual level.

In short, it includes everything you can learn about a person. Homeopaths should therefore be able to develop a truly individual therapeutic approach, more so than therapists employing other methods.

In my opinion, this is the second advantage of homeopathy: it aims to provide the patient with individualized therapy after a holistic assessment of the condition. Originally, this was to be done through individually selected drugs. Today we must realise that this is obsolete. But the idea of individual therapy is more topical than ever.

4.6.1 Homeopaths as Possible Health Coordinators

A well-trained homeopath could act as a kind of health coordinator and decide which therapy method or methods are appropriate (and desirable) *for a given patient*. This could of course also include conventional medical measures. In some other cases, however, the homeopathic consultation alone could be decisive.

The kind of conversation which homeopathy offers is already likely to act as a kind of therapy on an emotional level, analogous to psychotherapeutic therapy. The patient feels accepted, valued, and cared for in an empathic way. There is sufficient time available for the patient to express himself/herself and describe all his/her personal worries, fears, feelings, and peculiarities. This can help to reduce stress on the emotional level. As a further form of homeopathic therapy, the "icing on the cake" as it were, it may be possible in some cases to detect an individual sensation and bring about self-recognition. This process can contribute to a change on the spiritual level of a disease, analogous to the methods of depth psychology. The focus here will be more on spiritual health. This form of therapy is mainly possible through the sensation method in homeopathy.

To return to the initial example, if Mrs. M. wishes to be cured on the physical level through homeopathy, this can only be done indirectly: in a first step, by reducing the emotional part of the disease through homeopathic conversation. This therapeutic approach can be supported by globules via a postulated "super placebo effect" ("I will help exactly *you* in *your* illness"). This does not require any material or "energetic" active ingredient. The placebo effect acts through spiritual imagination. If it also includes precursory self-recognition of a spiritual core sensation, it may have an even greater effect and trigger life changes.

However, if Mrs. M.'s focus is on the desire for a purely physical improvement, the homeopath can initiate other specific measures. This could be a pain relief therapy, even using conventional medication. Recommendations such as physiotherapy, yoga, or sports therapy would also correspond to my image of the homeopaths as a health coordinator. In contrast to the general practitioner, who is theoretically responsible for this task, the homeopath can usually deduct his time expenditure appropriately (i.e., it is not included in the rigid fee system of the health insurance) - thus he has much more time available to meet the needs of each patient individually and holistically.

For our patients' benefit, we should take their sensations seriously and combine the advantages of both (and other) methods in a meaningful way. For physical problems, for example, we can choose conventional medical methods, and for other levels, for example, homeopathic conversation therapy.

In Table 4.2 I make a proposal for treatment options at the different levels.

I'm also thinking of some of my patients who had cancer diagnosis and were thus confronted with death. With scientific medicine, they have little chance of talking about the fear accompanying this. They'll get good surgery, chemotherapy, or radiation options. The physical area is covered in the best way possible for such a serious diagnosis today. But where can they turn with their emotions and essential questions? Here, homeopathy with its special setting can help to create a space in which patients can also express these complaints (psychotherapy is not always provided at the same time, since there is no psychological diagnosis, "only" cancer). The homeopathic conversation may lead to relaxation (which can be enhanced by the administration of globules), and in the best case this will also have a health-promoting effect on the physical level. It may also initiate individual life changes. A change in diet, an accompanying phytotherapeutic agent, or something similar can be used to support the measures required by conventional medicine. This results in a synergy between the two methods which will help the patient to regain health. However, such synergy can also help patients to face up to the situation when they reach the end of the options available with conventional therapy. We can also use the open setting of the conversation with the homeopath for this purpose.

All in all, the objective should be to encounter the patient on a fully human level and to accompany him/her medically and competently through every

Table 4.2 Levels of a disease and treatment options

Physical level	Scientific medicine, phytotherapy (herbal therapy), manual therapy, change of diet, sports, etc.
Emotional level	Psychotherapy, psychosomatic medicine, homeopathy as conversation therapy, self-help groups, etc.
Spiritual level	Homeopathy as consciousness therapy, awareness-building, meditation, psychotherapy, systemic therapy, etc.

aspect of human existence - on the physical, emotional, and spiritual levels. In case of doubt, it would not be necessary to use globules (which are only a kind of carrier for this principle), but rather the images of the human being and the disease anchored in the homeopathic method. And even though this broad picture may initially seem unusual for scientific medicine, according to its own WHO definition, it actually uses the same framework as the homeopathic holistic approach and so is not far from it.

I am aware that I'm describing an ideal which could not exist in reality. In homeopathy there is little clarity about the basic concepts, so it is not obvious that any serious health coordination would be possible at the present time. A lot of clarification and education would be necessary here. Training would have to be restructured, too, and perhaps even the conditions for admission to training courses.

There are two completely non-homeopathic reasons why the homeopathic method could be interesting for normal medicine:

- It helps us to decide at which level (or levels - they do not contradict each other) the patient needs help. This does not always have to be the physical level, even if there are physical symptoms. However, it also means that an actual physical disease (e.g., bacterial cystitis) can be treated physically (materially), e.g., by administering an antibiotic.
- Many unnecessary and costly examinations can be avoided by serious and competent decisions accompanied by patient education. A spiritual and/or emotional problem can't be detected by an X-ray or cardiac catheter examination and it can't be treated by an antibiotic.

4.6.2 The Life-Changing Goal of Self-knowledge

However, we have seen that there are some problematic concepts in Hahnemann's homeopathy which cannot remain unchanged if we want to continue to use homeopathy today. Medical and scientific developments require new definitions (and a bitter farewell) here.

- I strongly disagree with the view that homeopathic medicines contain something spiritual, ethereal, or energetic (i.e., something like "the essence" of the original substance) or information. We must say farewell to this view. The globules contain nothing. They are ineffective as medical substance.

- Hahnemann's idea that there is a vital force (i.e., something superordinate to the physical body) I can accept as an idea, but not as a fact. We may contemplate such an idea, but not as part of medicine. Despite all efforts, no equivalent of such a force has been found in the human body, and it would in any case contradict physics and biology. Physicists would want to award a Nobel Prize to any homeopath who found such a thing.
- The term "spirit" needs to be precisely defined before we can deal with it in medicine (as a possibility or a suggestion).

Human evolution has allowed us to develop complex ideas and perceptions and to experience differentiated sensations. Ideas and spiritual perceptions work in us ("I can design a large building that is safe for those using it", "I have energy", "I feel trapped"). Ideas can inspire, inhibit, block, and change. Let's assume that there is an idea in homeopathy of an "energy" or a "vital force" working in humans. In this sense, "energy" can be a figment of spiritual imagination, an idea. If it is a spiritual concept, there's no need to make efforts on the physical level at this point. (However, there isn't necessarily a contradiction either, since basically the same laws apply; I have explained the principle of emergence in the section "The Homeopathic Disease Pattern"). Let's assume that in homeopathy there is an idea that a vital force can be disturbed, and that certain information or ideas are required to let this vital force flow undisturbed. If we regard these terms as spiritual concepts, we will not come into conflict with science!

I use the terms energy, information, idea, vital force, etc., here only in the sense of ideas, such as were necessary in Hahnemann's time to interpret phenomena that we can explain much better today as part of science. We therefore no longer need to consider them as facts.

Hahnemann thought, however, that the ideal of an undisturbed flow of vital force could only be restored with the help of "similar" medicines, which in addition would have to be potentiated, i.e., spirit-like. When we bear in mind the redefinitions listed above, this can be interpreted as meaning that, whenever a disease can be described as a spiritual problem, i.e., at level three, it can be influenced by spiritual imagination (an idea or information). In fact, in this case, a material medicine is not always needed for healing. Healing may happen "by itself" if something is altered that hitherto hindered health changes. This can be done on a physical, emotional, or spiritual level, depending on where the therapist finds the patient's core problem. The only problem is to identify what kind of spiritual imagination that would be and how it could change anything.

Our vital force, as a spirit-like dynamis, cannot be attacked and affected by injurious influences on the healthy organism (…) otherwise than in a spirit-like (dynamic) way, and in like manner, all such morbid derangements (diseases) cannot be removed from it by the physician in any other way than by the spirit-like (dynamic, virtual) alterative powers of the serviceable medicines acting upon our spirit-like vital force, (…) so that it is only by their dynamic action on the vital force that remedies are able to re-establish and do actually re-establish health and vital harmony (…). (Hahnemann 1997, Organon, Paragraph 16)

If we see the core problem in spiritual imagination or sensation (which may be expressed retroactively at the symptom level), then, according to Hahnemann's idea, a spiritual conception is needed to solve the problem. This spiritual idea is worked out in conversation with the homeopath, and it can lead to a new self-perception: "Oh, that's how it is with me." In our introductory example, this was the moment the patient realized: "I not only feel symptomatically stiff and cramped, but this has something to do with my overall feeling!" Now we have raised awareness of a previously unconscious process which stressed the patient generally and appeared in the symptom.

While researching the theme of sensation outside of homeopathy, I came across an interesting psychotherapeutic method that seems relevant in this respect. The American psychotherapist Eugene T. Gendlin (a member of the research team led by Carl A. Rogers, who founded person-centred and humanistic psychotherapy) investigated those cases in which psychotherapy shows particularly good results. He came to the following conclusion: what was decisive for success was not so much what psychotherapeutic method was used or what the patient said during the session, but *how* he said it. If the patient was able to *feel an immediate physical experience while speaking* and was able to talk about it, deep psychological problems could be solved. Gendlin called this technique "*focusing*" and the physical sensation "*felt sense*". When this is achieved, Gendlin reported, there is immediate relief and an acute raising of awareness (Gendlin 1978). In the case of sensation, he seems to have made similar findings to homeopathy.

I consider the process of self-knowledge to be crucial in homeopathic therapy (as in other therapies, Fromm 1989; Mitscherlich 1974). Only a psychological explanation seems to me so far sufficient for such an effect. We are all unconsciously shaped by our lifelong impressions and experiences. This is well known and accepted, at least within the social sciences. Our behaviour is thus largely predetermined (e.g., Fromm 1989; Mitscherlich 1974; Precht 2007). As long as we don't learn anything new, our feelings and behaviour follow well-worn "pre-programmed" paths. However, if a good homeopathic conversation leads to new insights about ourselves, this could be a reason for far-reaching changes. This conversation can indeed favour self-help (and also the often mentioned self-healing) without contradicting science. The homeopathic perspective here coincides with that of psychosomatics and psychotherapy. (Wikipedia; keyword Psychosomatic medicine; Uexküll 1977). In 1974, Mitscherlich[8] wrote:

> As therapy (the psychosomatic method) is effective if, with the help of linguistic understanding, it succeeds in allowing the patient's fantasies, carried by affects, to advance into the vicinity of the conscious ego, and at the same time stabilize and give open-mindedness to the conscious ego so that it is able to tolerate these fantasies, cope with them (and change them), rather than repelling them through a head-in-the-sand policy. (Mitscherlich 1974)

In theory, homeopathy offers just such an approach when implemented more or less seriously. In medicine, however, this would mean broadening the area of research: in addition to the physical-functional level, the emotional and spiritual dimensions of the human being would also play a role.

References

Fromm E (1989) Gesamtausgabe in 10 Bänden. Ed. Rainer Funk. Deutscher Taschenbuch Verlag, München (in German) – see also: Erich Fromm online, Index of works and E-Publications, also in English http://fromm-online.org/en/werke-von-erich-fromm/

Gendlin E (1978) Focusing. Everest House, London

Grabia S, Ernst E (2003) Homeopathic aggravations: a systematic review of randomised, placebo-controlled clinical trials. Homeopathy 92 (2):92–98. https://www.sciencedirect.com/science/article/pii/S1475491603000079?via%3Dihub. Accessed 20 Jan 2018

Hahnemann S (1997) Organon of medicine, 6th edn (Translation Boericke). http://www.homeopathyhome.com/reference/organon/organon.html. Accessed 20 April 2018

---

[8]https://en.wikipedia.org/wiki/Alexander_Mitscherlich_(psychologist).

Hahnemann S (2016) The chronic diseases; their specific nature and homeopathic treatment. Leopold Classic Library, Victoria AU

Koerfer A et al (2008) Training und Prüfung kommunikativer Kompetenz. Aus- und Fortbildungskonzepte zur ärztlichen Gesprächsführung. (Training and testing of communicative competence. Basic and advanced training concepts for conducting medical conversations). http://www.gespraechsforschung-ozs.de/heft2008/heft2008.html. Accessed 12 April 2017

Kohnen N (2007) Kulturphänomene: Die Botschaft hinter den Symptomen. Umgang mit fremdländischen Patienten (Cultural phenomena: The message behind the symptoms. Dealing with foreign patients). Hautnah Dermatologie 1:20–23 (in German)

Metzinger T (2013) Spiritualität und intellektuelle Redlichkeit. Ein Versuch. (Spirituality and intellectual honesty. An attempt). Selbstverlag, Mainz (in German) http://www.philosophie.uni-mainz.de/Dateien/Metzinger_SIR_2013.pdf. Accessed 18 April 2018

Mitscherlich A (1974) Krankheit als Konflikt. Studien zur psychosomatischen Medizin. (Disease as conflict. Studies in psychosomatic medicine.) Edition Suhrkamp, Frankfurt/Main (in German) – see also: The Oxford Handbook of Health Psychology, Chapters "From Pathology as Psychic Conflict to the Specifity of the Symptom" and "Expression of the Emotions and Illness". Oxford University Press, USA

Nowak P (2010) Das Gespräch zwischen Arzt und Patient: Zentraler Ansatzpunkt oder Stolperstein für ein "gesundes" Gesundheitswesen? (The conversation between doctor and patient: central starting point or stumbling block for a "healthy" health care system?) (in German) www.patientenanwalt.com/download/Expertenletter/Patient/Das_Gespraech_zwischen_Arzt_und_Patient_Dr_Peter_Nowak_Expertenletter_Patient.pdf.pdf. Accessed 29 Mar 2018

Precht RD (2007) Who am i and if so how many? A journey through your mind. Constable, 1st edn (2011)

Rogers CA (1942) Counselling and psychotherapy. Houghton Mifflin Company, Boston, S. 123. https://archive.org/stream/counselingandpsy029048mbp#page/n147/mode/2up/search/conflicts+of+the+individual. Accessed 28 Mar 2018

Sankaran R (2003) An insight into plants, vol. 1 (Anacardiaceae). Homeopathic Medical Publishers, Mumbai

Sankaran R (2005) Sensation refined. Homeopathic Medical Publishers, Mumbai

Sankaran R (2009) The other song. Homeopathic Medical Publishers, Mumbai

Schmidt-Salomon M (2014) Hoffnung Mensch. Eine bessere Welt ist möglich. (Hope on man. A better world is possible). 2nd edn. Piper, München (in German)

Shang A, Egger M, et al (2005) Are the clinical effects of homeopathy placebo effects? Comparative study of placebo-controlled trials of homeopathy and allopathy. Lancet 366(9487):726–732

The Lancet: the end of homeopathy (2005) The Lancet:366(9487):690, 27 August 2005 DOI:https://doi.org/10.1016/S0140-6736(05)67149-8

von Uexküll T (1997) Psychosomatic medicine. Lippincott Williams and Wilkins, Philadelphia

Weymayr C, Heißmann N (2012) Die Homöopathie-Lüge: So gefährlich ist die Lehre von den weißen Kügelchen. (The homeopathic lie: that's how dangerous the theory of the white globules is). Piper, München (in German)

5

What Remains of Homeopathy in the 21st Century?

There are many reasons why patients turn to a homeopathic therapy. I described these in the previous chapters. If we take into account recent knowledge of those parts of homeopathy we can no longer accept from a scientific point of view, a new picture of homeopathy emerges.

Fundamentally, there is no such thing as homeopathy, in the sense that there is no single definition for it. It is therefore difficult to make an overall assessment of the method. My assumptions therefore refer above all to the form of homeopathy that I have described, which essentially goes back directly to Hahnemann's writings. For some points, I also refer to the sensation method which has been developed from Hahnemann's texts. I have already explained the reasons for this and the procedure involved there.

5.1 Which Parts of Homeopathy Are to Be Discarded?

According to the considerations described above, homeopathy can no longer be considered a medicinal (drug) therapy. The theory of similarity, vital force, and the traditional ways of making globules (potentiation) cannot be substantiated by a modern scientific approach, particularly with regard to possible effects. I completely reject this part of Hahnemann's theory. However, taking the globules may cause a placebo effect. I will go further into this below.

The drug pictures that have been built up in drug trials are untenable; they do not establish causality. The theory of homeopathic drug testing is scientifically untenable and can also be rejected.

© Springer Nature Switzerland AG 2019
N. Grams, *Homeopathy Reconsidered*,
https://doi.org/10.1007/978-3-030-00509-2_5

Another moot point concerns talk of a "subtle" (or "ethereal") energy, which is supposed to lead to illness in the case of malfunction and to healing in the case of correction. From a scientific point of view, such "energy" doesn't exist - neither in the patient, nor in the globules. Today, "energy" is a clearly defined physical term with the meaning "capacity to perform work". So not only is the term wrongly chosen against the background of present day knowledge, even the mere assumption that there is such an energy in the human body and in human existence is also wrong. I have already explained why. We must completely reject this part of Hahnemann's method, if homeopathy is to be applicable in medicine and hence in science in the future. A vital force can still be imagined today, but it is not a fact or even a medical quantity.

5.2   Which Parts of Homeopathy Need to Be Reconsidered?

I have shown how homeopathy works as a kind of "talk therapy" which aims to approach the complexity of the human being. Hahnemann was certainly ahead of his time with regard to this feature, and even today it is worth reconsidering some of his ideas.

In contrast to modern scientific medicine, the concept of homeopathy tries to portray the patient as a human being, rather than focusing on a mere symptom. It attempts to include the person's spiritual level, as well as the emotional and physical levels - both in the theory of disease development and in the intended healing process. Because homeopathy tries to cover and treat the physical, emotional, and spiritual aspects, we speak of a holistic approach. I have described how homeopathy proceeds here to give justice to this concept, although I am fully aware that are many different forms of homeopathy and that it is not a universal approach. This notion of a holistic approach distinguishes homeopathy from scientific medicine, which places purely physical symptoms in the foreground, and from psychotherapy, which focuses on emotional problems. This holistic approach obviously still makes homeopathy interesting for many patients. But if we reconsider this seriously, "holistic" also explicitly means including the physical level with its measurable and assessable findings. The findings of physical examinations concerning blood types, blood pressure, ECG or CT scans, etc., are also part of a holistic picture, but unfortunately these are often neglected in homeopathy.

As I have already said, I do not adopt Hahnemann's principle of similarity in the sense of a similar energy (as in vital force and homeopathic medicines).

However, with the homeopathic method, and in particular the homeopathic conversations, something can be recognised in the patient that gives a coherent overall picture. This would have to be concretized in dialogue. Hardly any critics of homeopathy doubt that the intensive conversations offered by homeopathy could have a positive effect on the patient. As a practitioner, I have tried to describe how homeopathy works (regardless of the problems due to the many different approaches within homeopathy) and where and how it can go beyond a normal good (general practitioner's) conversation. Through time, empathy, devotion, and the individual and holistic approach of homeopathy, there is more space to express personal needs and wants than in our poorly reformed everyday health system. The homeopathic kind of therapeutic conversation adds the dimensions of emotional and spiritual[1] problems to the patient-therapist contact and thus differs from a normal good conversation (or the purely fact-finding inquiries of some homeopaths). The methodology is similar in many ways to the psychological communication patterns of active listening, open questioning, and, in the case of sensation, paraphrasing and targeted elaboration of previously unconscious features. I have explained this in detail in the relevant chapters.

In the best case, patients can then rethink their situation and actively do something for themselves and their health (self-healing). Furthermore, they can dispense with conventional medical drugs where they are not absolutely necessary and thus also avoid their side effects. (The fear of side effects is frequently expressed by patients who turn to homeopathy - whether justified or not.)

Patients seem to have been helped so far by the thought that homeopathy recognizes a "picture" in its particularly symptoms and sensations and that it has a suitable medication for that. Placebo effects occur here, and these can be intensified by the deep doctor-patient relationship. Such factors contribute to the patients feeling well looked after and emotionally accepted in homeopathy. The core sensation elaborated and developed in the sensation method

---

[1]The term "spiritual" is a difficult term that is often misunderstood and misused. I use "spiritual" here only in the sense of the following definition, which does not contradict natural science, but which is certainly still unfamiliar as part of medicine (see the section entitled "The terms 'spirit' and 'spiritual' "). The spirit of man is the creative instance of consciousness that is superior to intellect (mind). This is where ideas, imagination, and suggestions are produced and put to work. A spiritual problem is one that arises not so much in reality as in the individual perception and imagination of the patient. We can surely *imagine* a vital force here, but this has nothing to do with reality. In turn, our imagination can be changed through further pieces of imagination, information, or suggestions (e.g., Berger et al. 2013).

Including the spiritual level does not mean spiritual healing or anything of that kind. With the term "spirit" I don't want to let some kind of irrational magic come in "through the back door". Involving the spiritual level means getting closer to the patient in his/her complexity as a human being, and this may have an influence on health that has been underestimated (or has been difficult to grasp) so far in scientific medicine.

is the essential component (the core) of the patient picture. In the past, such a patient picture has been juxtaposed with physiologically ineffective drugs on the intellectual level of ideas, which in turn can only work through ideas. (For example, a patient can have the idea: "I experience a feeling of being trapped and restricted", and receive the information that the medicine contains something that can change this idea.) It is doubtful whether this will continue to work effectively in the future when patients discover that their medicines have only a non-material significance.

The therapy consists first of the discussion in the common elaboration of the patient picture on the spiritual level, and can subsequently be supported by globules. I see the globules as a carrier of a suggestive power. When taking the globules, the patient becomes generally aware on the emotional level: "This will help me" and "I want to follow this way of healing right up to the goal". On the spiritual level, this effect more specifically and purposefully combines with the core sensation or some erroneous notion. The globules can thus become the carrier of an even stronger autosuggestion ("This will help *me* with *my* complaints"). The globules serve as a "reminder" of the homeopathic setting and the jointly elaborated core sensation. They are substantively (materially) ineffective, but important carriers of the information necessary for change on the spiritual level.

According to these considerations, the pure reminder function of the globules is decisive, not the way they are made or their supposed content. It would henceforth make no difference whether the globules "contain" Natrium muriaticum or Sepia. Basically, this has never mattered, because they all contain the same thing anyway: nothing. One could therefore use pure lactose globules (or other carriers of autosuggestion).

The effect of globules may be supported by the often practised secrecy of the drug. This sometimes encourages the patient's self-observation and self-initiative; they must watch out for what is triggered by the drug intake, pay more attention on their own, and as a result they will learn to take better care of themselves. From placebo studies in scientific medicine, it is also known that exact recommendations for use (as detailed as possible) and special dosage forms lead to particularly large placebo effects (Wikipedia, keyword placebo). Both conditions would be fulfilled in homeopathy.

The effects on the emotional and spiritual level could indirectly have a positive effect on basic physical processes, complaints, and symptoms. I will go into this in detail in the next section. Direct physical or physiological effects do not appear. (Hahnemann never claimed that either. He assumed that the disorder of the vital force was expressed through symptoms. Only by a change in the vital force could the symptoms be made to improve retroactively.)

5.3 Why Should We Think Again About These Points?

Normal medicine cannot yet make deeper inroads into the complex human system. In homeopathy this is at least attempted. If we want to prevent patients from migrating to alternatives - some of which are practised in an irresponsible way - we must create a basis and offer a procedure for dealing seriously with these patients. This can be done either within some reconsidered form of homeopathy or by (re-)integrating parts of homeopathy into everyday medicine. Of course, these possibilities are also available outside of homeopathy, in psychotherapy or psychosomatic medicine. In particular, the systematics of medical cybernetics deals with the complexity of human beings in medicine:

> Medical cybernetics is a branch of cybernetics [...] which applies the concepts of cybernetics to medical research and practice. It covers an emerging working program for the application of systems and communications theory, connectionism, and decision theory on biomedical research and health-related questions. [...] Medical cybernetics searches for quantitative descriptions of biological dynamics. It investigates intercausal networks in human biology, medical decision-making and information-processing structures in the living organism. (Wikipedia, keyword: Medical cybernetics)

In the Penta-model due to *Foerster and Burrer* I found similar approaches to those suggested by homeopathy. Besides the emotional and spiritual level, the social level is also included in this model, and the physical level is divided into biological and physical influences:

> According to the control loop interrelationships between nature and society, the principle of interaction also applies to medicine. In such an interaction, physical, emotional, mental, and social areas and the ecological (biological, physical) environment influence each other in man and cause an internal as well as an external process of adaptation. Fundamentally, the environment, body, and soul touch and steer each other mutually. The autonomy of a patient, his or her social integration as well as mental and emotional influences must therefore be given special consideration and therapeutically controlled. (Burrer 2013)

The research is avowedly evidence-based, which would be an advantage over homeopathy as it has been practised so far.

However, there is still a long way to go before these theories can be implemented in everyday medical practice. Psychology is already more advanced. However, this path is much more difficult for patients. It stigmatizes ("I haven't lost my marbles!" "What will my employer say about this?"), leads to difficulties for health insurance, and requires a psychological or even psychiatric diagnosis. A visit to the homeopath (especially if it is recommended by a best friend) or a homeopathic self-treatment seem much more obvious. This free access is another important part of the good feeling that homeopathy offers patients and should not be underestimated.

The question is: do we really need homeopathy today for the points mentioned here? And all the more so in that the principle of similarity must be considered obsolete, whence homeopathy should not even be called that any more. In the next step towards an answer to this, we should look again at the facts and what science has to say about them.

5.4   How Can We Take a Stance on This Through Science?

In the last section I explained my assumptions about how the remaining contents of Hahnemann's system of ideas, i.e., those features of homeopathy as a method that are still worth considering, might be reconsidered in the 21st century. I have drawn these conclusions, as described in this book, from my own dealings with homeopathy and my patients and from my analysis of the theories of the original homeopathic doctrine. These conclusions are now available for review, which is my intention in this section. However, I would like to point out at the outset that the discussion now becomes more scientific (for homeopathic conditions). This section is aimed particularly at medical professionals and readers interested in scientific research.

There is no way that we can carry out a full scientific investigation of the reasons why patients feel attracted to homeopathy and feel cared for when they appeal to this approach. A "good feeling" has too many indeterminable factors whose influences cannot be precisely quantified, not least because they differ from one individual to another.

The basic question of "Does homeopathy as a method have any effect?" can therefore only be divided into sub-areas:

- Physical level: Direct changes in a physical symptom or indirect changes indicated by a measured value could be assessed according to the standards of medical studies.
- Emotional and spiritual levels: More complex changes on the emotional and spiritual levels are more likely be picked up in psychological studies.

- Another suggestion: The question of whether a targeted effect, perhaps only a super-placebo effect, may occur if the globules are linked to an individual message could also be investigated in medical studies.

Let us begin with a short digression about how to proceed with such a review in scientific medicine.

How to test the effect of a medicine?

Before the introduction of the modern principle of evidence-based research, the procedure, simply explained, was as follows: what changed in a patient after taking a drug was observed and these changes were recorded. They were almost regarded as proof of the efficacy of the drug. But the problem is this: it was not possible to prove that the changes observed were caused solely and absolutely by ingestion of the given drug. Other causes may have played a role (e.g., changed living conditions), or indeed the complaints might have gone away of their own accord during the observation period. There was no definite association between cause and effect, no obvious and logical causal connection (which is the basis of evidence-based research today). With this approach there was no attempt to record why and how often there were *no changes* after ingestion, which would therefore have been *exceptions* to the (expected) rule, and no attempt to understand how they came about.

In homeopathy, research has so far worked like this: one patient took the globules and the general condition or physical symptoms changed, so it was assumed that this change had been caused by the intake of the globules. However, it would be impossible to prove this on the basis of this result alone. Too many other factors may have played a role. Moreover, we homeopaths can't even explain how this could have been happened because, as we've shown, our globules are not based on a logically comprehensible mechanism, and we are no longer willing to accept other explanations.

Science works in a different way. We see that something is changing and try to find out which (known and logical) principle could be behind it and whether the postulated chain of cause and effect can be reproduced. However, there is an obvious connection if we can show, not only that A leads to B, but also B does *not* occur if A is omitted. This is strictly essential for being able to speak of causality.

Here is an illustration. A mother gives Arnica globules to her child after a fall and observes that the bruise disappears quickly. The next time the child falls, she would have to observe what happens if she doesn't give any globules. Furthermore, she would have to observe about one hundred other children who receive Arnica after a bruise - or not - and check the results to see whether there is a connection that goes significantly beyond expectable coincidence and/or whether there may have been other influences that put the result into perspective (e.g., severity of falls, age of children, type of clothing, etc.).

She should go even further. For example, some mothers who have experienced in the past that their children react particularly strongly to bruises strongly may fall back on Arnica - and others may not. Then we would see two groups with different characteristics - with different sensitivity to bruises - and these characteristics might have an impact on the result. The children to be observed should therefore be divided randomly into two groups.

We choose a hundred children and roll a dice to decide which child gets Arnica and which doesn't. The placebo effect, which could arise from the fact that a child is recognizably given a (supposed) drug, must be excluded. All of them would have to get globules; some pure placebo and others with Arnica. Children and parents are not allowed to know who receives what and - if possible - neither should the person evaluating the data. Only at the end, when all the children have been evaluated (for example, using a score system), would the evaluator reveal which child has received a placebo and which Arnica. This would be the current gold standard of medical research: a double-blind placebo-controlled comparative study.

Of course, no mother would go that far; she would rely on her feelings, her experience, and perhaps the feelings of other mothers. But this has nothing to do with scientific procedures. The above outlined effort of checking (and falsifying[2]) the hypothesis "Arnica has a significant influence on the development of bruises" is taken over by evidence-based research.

---

[2]Falsification is the generally accepted scientific method used today to test a hypothesis (in our case, the hypothesis that homeopathic remedies have a medical effect), i.e., to check their truthfulness. If a hypothesis can withstand this, it is regarded as a preliminary state of knowledge until such times as scientific progress may later refute or modify it. If a hypothesis cannot withstand the test, and if there are no systematic errors and misunderstandings in it that could be eliminated, then it must be regarded as refuted. An "absolute truth" is not known in the principle of falsificationism, only approximations to it in the sense of statements of probability. In critical rationalism (Karl Popper), this is the maximum attainable state of knowledge.

Neither is it possible in this sense to prove causal connections in any definitive manner (such an attempt would lead to an infinite regress). However, unlike verification, falsification allows conclusions to be drawn about the degree of probability of causal relationships. This is precisely why falsification has established itself as a standard in (medical) research.

Obviousness and justifiability of cause and effect are therefore imperative to exclude the possibility that A and B follow each other only by chance - and that goes for homeopathy, too.

That's why I deliberately wrote at the very beginning of the book that I have seen this or that disease disappear *under* homeopathic therapy. For it is quite simply not possible to say with certainty whether these changes have occurred *as a result of* homeopathy. Homeopathy can't be substantiated by its unscientific part, i.e., the part that assumes any medical effect of the globules (via energy and vital force) to be a homeopathic effect.

So far, all studies on homeopathy carried out correctly according to the procedure described above have only ever shown this: it makes no difference whether we give Arnica globules or empty globules. The effect is the same in both cases. The effect corresponds exactly to the effect that occurs when someone receives any alleged drug that he or she believes will help him or her: the placebo effect.

All this and the lack of explanation for any medical effect in homeopathy must lead us to the conclusion that homeopathy is not a drug therapy. At best, it has a placebo effect and is otherwise limited to a kind of conversation therapy. I have tried to demonstrate the principles of homeopathy as a method: how it works through the doctor-patient conversation, and what distinguishes it from other forms of conversation between therapist and patient. Now the time has come to put this method to the test. However, the above-mentioned evidence-based research principle becomes more complex if one wants to test an entire method rather than a single drug. The many incalculable influences and uncertainties make reliable assessment difficult. But if we go by the principle of falsification and run some studies (viz., the new scheme discussed below), we will be able to say with near certainty whether homeopathy as a method has a proven influence or not.

> The highest ideal of cure is rapid, gentle and permanent restoration of the health, or removal and annihilation of the disease in its whole extent, in the shortest, most reliable, and most harmless way, on easily comprehensible principles. Hahnemann (1997), Organon, Paragraph 2

That's what Hahnemann himself says about healing. However, he himself and all his successors have so far failed to provide these "easily comprehensible principles". I therefore propose a new procedure to check whether there is any truth in homeopathy:

- We accept (as homeopaths with a heavy heart) that there is nothing in the homeopathic medicines that could explain the effect, let alone an energy or a spirit. We don't need to do any further research on this.
- We separate the drugs from the method. They are substantially (materially) ineffective, but can act through a placebo effect (which we could investigate further in this context). Homeopathy is no longer a drug therapy.
- We examine whether, as a kind of conversation therapy, the homeopathic method has effects on the emotional and spiritual level or - retroactively - a measurable influence on underlying body processes. Here we cannot test homeopathy as a whole and must first turn to individual questions instead. Of course, it is important to substantiate that homeopathy in this form is equivalent or even superior to other, already established methods. But that is yet to be proven.[3]

For homeopathy with globules this means in concrete terms:

We must first acknowledge that claims about the effects of globules are unjustified. Nothing works here except for a placebo effect. We would have to find evidence for this placebo effect too, but we could at least justify it (possibly even as a "targeted", individualised placebo effect). Physiological changes are definitely possible as part of a placebo event. But then we must recognise that the impacts of a placebo will be difficult to measure directly due to their great complexity. From spiritual to emotional and physical changes, everything is possible. We could link the change to concrete physical symptoms and/or the core sensation. For example, the research question could be: Does the placebo effect in homeopathy go beyond the normal placebo effect or not?

In the last section, for homeopathy as a method, I described what the *causes for the possible effects* of this method could be and at what level they could be beneficial to people, i.e., where and how these effects could unfold. Accordingly, there are (complex) reasons for such effects. However, the *evaluation of these effects* is much more difficult, for the following reasons.

If homeopathy is to become part of medicine and thus part of natural science, it must not contradict well founded and well researched scientific laws. However, an approach based *purely* on natural science would not lead to

---

[3]A study by Brien et al. (2010) is already taking this approach and concludes that homeopathic consultation has a positive effect (in contrast to the gift of globules).

the desired result, since homeopathy also aims to encompass complex spiritual processes. The spiritual level of a person is an unusual notion in medicine. The step to esoteric spiritual healing and charlatanism is short and we may find ourselves dangerously close to the unscientific (from which we homeopaths must free ourselves, as I have explained). I have taken the spiritual part of the human being as given - knowing full well that this touches the areas of spirituality, philosophy, psychology, and even esotericism, which I cannot deal with exhaustively in this context; these aspects have not yet become part of medicine. The human spirit might be relevant to medicine if it became clear that many diseases originate in the spiritual realm, or are at least modulated by it in some ways (as homeopathy suggests). Up to now, medicine has been limited to the physical sphere and has achieved outstanding results in this area. Here, medicine is measurable, here medicine takes place in the area that can be recorded scientifically, and here studies can yield unambiguous results.

Studies in the emotional and spiritual fields are much more difficult. Which criteria should be used for a reliable assessment? Due to the complexity of the human being, the scientific paradigm of separating cause and effect has its limits, especially in the emotional and spiritual spheres, where not all influencing factors can be expressed in numbers and thus made accessible to statistical analysis. Here, multi-causality is the norm, and it is not as easy to judge as a one-dimensional cause-effect principle. But why should medicine be limited to one level? I am convinced that the spirit *could* come part of medicine. It is important to find ways for scientific research to go beyond the physical level and gain access to the emotional and spiritual levels. From the theory of emergence we know that the interaction of several summands can result in more than the mere sum because synergies occur, for example. A system can develop properties that do not result on a one-to-one basis from the individual parts. The interaction of fundamental neurophysiological and biochemical processes results in our consciousness, our thoughts, and our feelings. These processes can be explained individually, but the qualitative leap to a "great whole" cannot be fully derived from physics and biology alone (bottom-up emergence). We may be able to use this the other way around (top-down emergence). Even if we cannot explain the "big picture" of homeopathy, we may be able to examine individual and basic parts of its effects in more detail.

Here, I would like to quote again from the German Wikipedia:

> Human thought contents (ideas, concepts) have emergence characteristics in relation to the neurological processes and psychological acts from which they arise.

It is therefore possible that, due to the phenomenon of emergence, the human spirit, which is itself difficult to grasp, may have an influence on human biology, chemistry, neurophysiology, and so on, but this influence can't be explained on a one-to-one basis from first principles.

A spiritual idea such as "I'll relax" can lead to a reduced release of stress hormones such as adrenaline and cortisol. This happens through very complex connections in the human body, all of which are coherent and comprehensible in themselves, but which only have this effect through their overall interaction. Cortisol releases can be caused and controlled by various mechanisms in the body. It doesn't necessarily require a spiritual imagination for this - but they can also be influenced by the spirit. This, in turn, should be indirectly measurable.[4]

Such a procedure is no longer purely scientific, but it doesn't contradict natural science either. Beginning from the emotional level, we are no longer dealing with purely physical, measurable, biological and physical facts. Homeopathy, as I have described it so far, doesn't lie exclusively within the boundaries of science; this is what is special about it. It is not just about the physical level, it is not just about drugs, it is also about intra- and interpersonal issues and therapeutic principles. This is the realm of the social sciences, not the natural sciences. Social sciences like psychology approach their research subject in a completely different way, simply because their subject is an entirely different one. For example, the person is considered in their full complexity, so not only physical functional processes play a role. Medicine has already accepted this in some fields, but not yet in others. Research in the field of salutogenesis, a field which is not primarily concerned with disease (and its causes), but with recovery and the possibilities for preserving health, shows that spiritual attitudes do have an influence on physical processes. For example, patients who deal optimistically with a disease produce significantly more immune cells. Patients who feel helpless, on the other hand, show an increase in a stress hormone that leads to a reduction of immune T-helper

---

[4]A little experiment on this principle: Close your eyes for a moment and say to yourself, "I'm exhaling." Remain like this for a moment, and then read on.

Close your eyes again and say to yourself, "I'm breathing out in a long and relaxed way." Do you notice a difference in the depth and length of exhalation, in your overall muscular tension, or something similar? And would such changes be measurable? Would the effect of the mental imagination ("long and relaxed") thus be objectifiable?

cells and the production of antibodies (Lorenz 2004). In this branch of research, it is assumed - as in homeopathy - that humans have at their disposal some possibilities for controlling their health and recovery which they themselves can activate or deactivate. Research in the field of homeopathy should therefore deal increasingly with the question of the effects it has on the body, as a method, and its measurable and assessable findings.

To examine these emergent changes on the physical level, caused for example by following some kind of homeopathic therapy, here are some ideas about how we could proceed.

The first step would be to expand the field of research. It should be accepted that the emotional and mental levels are as important as the bodily physical level, and also that there can be repercussions between the levels. To some extent, the term "human sciences" has already been coined for this (Wikipedia, keyword: Human science).

Next, what we wish to examine must be presented as a falsifiable statement, e.g., the claim that the therapeutic setting of homeopathy reduces the allergy symptoms of patients. Then we need to specify which data is to be used for assessment. Which patients, which diagnosis, and which criteria (e.g., typical symptoms such as sore eyes, runny nose, and sneezing, but also typical measurement values such as immunoglobulin E or histamine) will be examined and factored in the study? For the purpose of testing the hypothesis, a comparison group with the same diagnosis and purely normal medical treatment would then have to be provided. Blind testing (no participant or therapist knows which group he or she belongs to) is not possible in this case, because the patient and therapist will notice whether they are in a homeopathic setting or not. The study is then designed, which means determining and illustrating how the study will be carried out in detail. This can be very difficult, as many possible influences must be taken into consideration before beginning. In our example, the duration of treatment will differ considerably: conventional medicine five minutes, homeopathy one and a half hours. Will the time factor (the duration) alone influence the measured values? Do different therapists or different homeopathic methods have an influence? How can we find out and neutralise these influences as far as possible to ensure that differences in outcome can be exclusively attributed to the different treatment situations?

The study will now be carried out and the data will be evaluated. The most important question is whether there are differences between the homeopathy group and the comparison group, and whether these differences could have arisen solely by chance. To rule this out, the first thing is to examine groups that are as large as possible, with randomly and variously combined patient cohorts. Secondly, the results obtained in this way must be checked by means statistical significance tests to determine whether they correspond to a random distribution or deviate from it by more than just chance. It is also important to know whether there have been any unforeseen events that could have had an impact on the outcome (e.g., a particularly mild allergy season; regional differences in pollen count; have patients in the homeopathy group not in fact taken conventional medicine?). All these factors must then be discussed and presented in a scientific publication.

Moreover, more complex influences of the therapeutic setting of homeopathy could be investigated in this way. While in every supermarket our perception is deliberately influenced by manipulations of scents and light, in medicine we neglect the sensation of the individual patient (as an expression of the spiritual part of the human being). In marketing, the influence of perception is measured indirectly by buying behaviour. What indirect criteria could we use in medicine to describe the influence of sensation on health? Perhaps subjective assessments of the patient's state of health before and after a (homeopathic) therapy, e.g., in the form of a scale. For example, in the case of illness, could a quantitative comparison of the intake of conventional medicines, e.g., reduction of painkillers, be an indirect indication of the patient's sensation? Could brain research prove the influence on sensations? Or could criteria such as blood pressure, stress hormone levels, immune parameters, etc., be exploited to draw conclusions?

An exemplary procedure similar to the study draft outlined above could go something like this. The falsifiable hypothesis could be: "The homeopathic setting reduces a patient's sense of stress." In this case, criteria for emergent physical change could be: daily cortisol profile, and the levels of hormones and hormone precursors that may be elevated under stress, like basic serotonin, adrenaline, noradrenaline, and DHEA (Henzen 2004; Kirschbaum 2001).

However, such a procedure will presumably be difficult because the individualised approach so characteristic of homeopathy always focuses on also individual changes. This was previously regarded as major argument against the possibility of being able to test the effectiveness of homeopathy experimentally. Such changes may be noticeable in personal conversations, but they are difficult to specify on the basis of objective criteria in studies. There are too many ways for such an approach to fail. Chance, the passage of time, and countless other factors of life such as moving to a new city, new partner, etc., may also have led to changes. Here I plead in favour of accepting the patient's specific (core) sensation as a "diagnosis" and primarily paying attention to changes here.

However, it must be borne in mind that such an effect and comprehensive healing may be more difficult to depict than a purely physical improvement in some bodily function. For example, it's easy to measure whether the lung function is restricted or not. But how can we represent in our studies a change in someone's life due to an insight, a change of job, a better attitude to life, or an improved self-perception? And in our introductory example, how can such studies depict what Mrs. M. summed up in the words: "The pain no longer plays a role in my life"? Or will that just not be possible, so that we will forever be unable to demonstrate our homeopathic experiences?

Studies with more comprehensive questions of this kind should be based on psychological or sociological study designs, rather than those of natural science. Tests are available for measuring well-being, for example the health questionnaire Short Form 36, a disease-independent measuring tool for health-related quality of life surveys. SF-36 is often used in medicine and psychology for therapy control or progress measurement. The scoring includes the following parametric sections:

- vitality
- physical functioning
- bodily pain
- general health perceptions
- physical role functioning
- emotional role functioning
- social role functioning
- mental health

(Wikipedia, keyword: SF-36)

In this area of psychological therapy, however, homeopathy as a method would first also have to prove that it is superior to or at least comparable to established psychological methods in terms of changes on the sensory level.

Let me now turn to the globules again. Reviews of the effectiveness of homeopathy have so far been tantamount to an examination of the effectiveness of homeopathic globules. After reading this book, perhaps it is clearer that this is ruled out from the outset because it is scientifically unthinkable. However, it is quite another idea to look more closely at the placebo influence of globule administration. If the globules contain general information ("I will help you") or better still very specific information ("I will help you with exactly your complaints"), and if this does indeed have a distinct influence on patients, could this not be proven in studies? For example, a conceivable procedure (again analogous to the studies described above) would be to form and compare three groups: (1) a homeopathic setting with the gift of completely normal globules and no further information; (2) the gift of globules with concrete information, e.g., linking it to the now conscious core sensation; and (3) information, but without globules, e.g., "Please think three times a day about the essence of our conversation and about what you want to change".

Another question concerning the effect of globules could be this: Does the often almost celebrated secrecy of the drug name in homeopathy make a difference (e.g., by promoting self-observation and thus a drive to practise more self-care)? Is it possible to find something by placebo research that is relevant to new therapeutic models, for example, the deliberate use of placebos? The question is, however, whether this can be in the interests of patients and modern medicine. To this extent, further research into homeopathy would at least remain honest: we admit that we only have placebos to offer and so carry out research on their spectrum of effects, and we research how helpful they may be, for example, in the context of self-medication.

All in all, it remains difficult to check the effectiveness of the homeopathic method. It remains difficult to translate into facts the "good feeling" of the patients which is so often emphasized. Nevertheless, I hope this book will provide ideas for further (evidence-based) research in this area. With the requirement of being a part of medicine and thus of natural science, it must and should be possible to present the effects of homeopathy conclusively and comprehensibly. Will we end up falsifying the hypothesis "Homeopathy as a method has no effect - neither on the emotional nor on the mental level nor retroactively on the physical level"? Patients should be able to rely on homeopathy to offer them something based on data and facts, not on personal opinions and beliefs, precisely because they themselves cannot check this, and ultimately must rely on their feelings. If we cannot offer this, we must reject the method.

5.5   Homeopathy as the Patient - A Last Example

Let's round off by personalizing homeopathy as a final example and seeing how it fares in the role of a "patient".

Figures, data, and facts suggest that homeopathy is sick. It lacks data to prove its efficacy, it suffers from postulates that can no longer be upheld today in science-based medicine, and it dreams up ideas of drug effects that nobody else can understand. That would be a description of the sick homeopathy at the physical level, so to speak, and these are her symptoms.

On the emotional level, homeopathy as a "patient" associates great fears with these symptoms, usually expressing them defensively, and refuses to accept arguments and facts. It prefers to counterattack ("There is something wrong with scientific medicine!").

On the spiritual level, homeopathy suffers from the delusion that science is not competent to judge it at all, that it lacks the proper methods to do so, that everyone is against it, and that nobody understands it. Homeopathy feels itself as misunderstood, condemned, and pilloried, although it only wants the best for its patients.

If our "patient", homeopathy, came to visit my practice with its symptoms, then as a homeopath I would not prescribe a painkiller to quickly remove the symptoms. Instead, I would try to learn more about it, and learn more from it. How do its symptoms relate to its feelings, sensations, thoughts, and daily actions? I would take my time, let it tell me how it is doing and listen empathetically until it has finished speaking. I would try not to condemn it outright or to find a quick remedy for it ("Just realize, homeopathy, that's all not true!"). I would like to listen to its genesis or history, understand its background and interrelationships, and try to work out a kind of individual guiding idea. What really is its problem? Where does the stress come from? What led it to me? Homeopathy would perhaps say something like: "No one understands me anyway. All I want is to do something good, and they're all against me. Luckily, I still have my patients and my loyal homeopaths. They give me their approval. So you see, it can't all be so bad!" A guiding idea or basic sensation could be: "I'm not at all sure whether I'm allowed to exist, but if I were ever to admit that, I'd be endangering my own existence." This sensation can be generalized. It applies to every single homeopath, as well as to homeopathy as a whole. It applies to the meagre facts of homeopathy as well as to their questionable overall presence in medicine - then just as it does now. We would thus have reached the core sensation. At the precise moment when homeopathy becomes aware of this, a short-term, deep relaxation would perhaps occur: "Oh, that's how it is with me! I fear for my very existence!" A deep self-knowledge would have been achieved.

Traditionally, as homeopaths, we would have given homeopathy a few globules that would (supposedly) have caused similar fears of existence in a healthy person. According to Hahnemann's theory, this would have enabled homeopathy to overcome its fear of existence.

Reconsidered, however, the following would probably have happened. Thanks to the therapeutic setting, the "patient" (homeopathy itself) would have felt emotionally well looked after and finally reached an understanding, whence its stress and resistance might have been alleviated by this alone. It could have turned to face reality feeling a little more relaxed. In addition, it would have learned more about itself through this self-knowledge: "I am afraid for my existence, because so far there is not much to support my *raison d'être*. I am now able to initiate a situation-appropriate self-change, to think in new ways, to face reality, and to try to deal with my deficiencies and symptoms in a new way." Maybe we will give it some globules, but with the explanation that they only act as a placebo, and at best contain the suggestion and carry the meaning that homeopathy should strive to identify its *raison d'être*. Whenever it doubts this or falls back on the old pattern, it should take the globules.

Now, a few weeks later, we invite homeopathy back to our practice and ask about the course of its disease. It reports with relief, but also in tears, that it has dealt with some facts and figures and now realizes that its previous views and meanings are hard to believe. It is certainly relieved, but also deeply shocked by this insight. At times it feels it can hardly continue along the same path. On one occasion it was attacked by severe fears, but it took the globules and remembered our conversation. Being able to do something positive about the situation was a help to it at that moment. Since then things have slowly improved. It has endeavoured to come up with new ideas for studies. Since it realized that it was acting so defensively only out of fear, it has led a much freer life. When it encounters scientific medicine or science, it is no longer so afraid, so doesn't feel that it has to fight it off so bitterly. Things are not perfect yet, but it's on the mend and it's getting along better with its symptoms.

In this sense, homeopathy as a patient has begun a process of self-healing, which should now be further supported. No law of science has been violated. It nevertheless remains to be decided whether this procedure can be part of today's medicine. It is an open question whether we can determine a change at the "number/data/fact level" by means of the principle of downward-acting emergence. This will take some time, and it will not be easy to find criteria that would allow us to conclude in the context of homeopathy that something has changed, not only internally (subjectively), but also externally (objectively, significantly). The changes may be complex. In personal conversation they are quite easy to grasp, but in studies? It will be difficult - perhaps impossible - to find clear parameters. But this seems to me more likely and more promising than further insistence that homeopathy is not sick at all.

5.6 What Now? A Conclusion

When I started working on this book a few years ago, I wanted to write a much more positive book about homeopathy. I wanted to clarify what homeopathy aims to do and what it can actually achieve. I started to get serious about the background for the first time. I read critical books and blogs, at first because I wanted to take a stand on them and invalidate their arguments. But my conclusion was that homeopathy as it stands is not medicine in today's sense. There's not much left when we consider all the facts. To begin with, there is no single well defined form of homeopathy, and too many conflicting currents make assessment difficult. In addition, the basic theories of homeopathy (i.e., those on which homeopaths at least agree among themselves, such as the manufacturing of medicines, drug testing, drug effects according to the principle of similarity) are not compatible with today's medicine, which is based on natural science and the principle of causality.

In a nutshell, homeopathy works because we as homeopaths and our patients have the idea that it works. Studies so far have not uncovered anything else.

We now have two options. One is to say goodbye to homeopathy as part of normal medicine. We will then no longer have to worry about scientific laws and we will not need any further studies. There are many esoteric movements with convinced followers; homeopathy could also find a place here. However, we could not then insist on reimbursement by medical insurance companies - either for therapy or for medication. The other is to remain part of medicine and say goodbye to the untenable aspects of our theory. However, it might then be more appropriate to assign homeopathy to psychology and psychological research, rather than to human science research, because it tries to include emotional and mental health, where different laws apply compared to those at the physical level. Whether homeopathy as a method has advantages over already established psychological methods, however, would first have to be investigated. Homeopathy would then no longer be a medical therapy and should no longer be called "homeopathy" (from the eponymous "similar cures similar").

One question has moved me throughout. Regardless of our thoughts about a reconsidered form of homeopathy and whether it could be confirmed in scientific research, we need to ask why there is such a sudden rise of interest in homeopathy, despite all the scientific criticism?

Homeopathy can (or would like) to do something that today's medicine is unable (or unwilling) to do. For example, it tries to see the patient as a human being and not as a symptom carrier. There should be no doubt that this doesn't always happen in a serious manner and according to the current state of knowledge. It is also clear, however, that patients quite rightly long for this. I still believe that this approach is an honourable one, and I fully agree with it as a medical doctor. It would be desirable to incorporate those aspects of homeopathy that make it so human into our rather perfunctory symptom-focused medicine. Since the introduction of Diagnosis Related Groups (DRGs) with a flat-rate billing procedure in Germany, and since time for treatment has been cut back, homeopathy has been booming. That speaks for itself. I could easily understand patients who came to my practice because they didn't want to be regarded as a machine with a functional failure. In the same way as someone relying on the high level of a current treatment guideline doesn't want to be squeezed into a rigid scheme that allows no room for individualization. The feelings and sensations of patients are important and they want someone who is competent in deciding which therapy is best suited and should therefore be chosen. The walk from the family doctor to the psychologist is still burdened with shame and guilt and often means long disputes with the health insurance company and problems in one's social environment. In addition, a psychological or even psychiatric diagnosis must be given as a prerequisite, which is not of course the case with most patients. Psychosomatic specialists are also psychologists and thus equally difficult to find. For patients, things usually stop here. If they go to their family doctor, medicine will be practised at the body level, and this is usually excellent, but still not sufficient for the patient as a human being. As a result, in (too) many cases, patients will turn to alternative medicine - the disadvantages are sufficiently well known, as described elsewhere in this book. People don't decide things solely on the basis of figures, data, and facts; they also like to rely on their intuition and feeling when making decisions - as they do when choosing a therapist.

So, if I had to sum everything up in one short statement, I would say this:

> Homeopathy is bad in theory, but good in practice, while the opposite is true in scientific medicine.

Scientific medicine offers facts and knowledge but approaches the patient as a human being in a rather rough-and-ready manner. As long normal medicine leaves people on their own to deal with their human feelings and sensations, even though these may be related to or lead to a physical illness, people will search for help in alternative fields. And we have no control over what happens there. We should not therefore denigrate alternative medicine and satisfy ourselves with denouncing its disadvantages, but instead take responsibility for what people hope to find there. This would be an opportunity to subject our health care system to a fundamental review, and with it our medical research programmes, where we might introduce a stronger connection with the humanities.

5.7 An Epilogue for Patients and Homeopaths

I'm aware that the toughest criticism of this book will probably come from your side. It wasn't easy for me to write it. I was absolutely convinced of homeopathy for years and even worked exclusively as a homeopath in a successful practice, with great feedback from my patients (although it may be that those less satisfied were just holding back). Things could easily have stayed that way. But then I started to deal with the background for homeopathy. What exactly happens during potentization? What are the actual results of the various studies? What do scientists say about the theory of homeopathy? How does homeopathic drug testing work? What did Hahnemann himself write, and when?

The results of my investigation shocked me - initially, of course, still paired with disbelief, but I became increasingly convinced that homeopathy was making us look bad. It's no easy matter to admit this.

> Anyone who has completed a long and possibly expensive education as a homeopath has a lot to lose. If a sceptic tries to convince such a person that globules are nothing but small sugar beads and that there is at best a placebo effect behind this kind of alternative medicine, he will not be impressed. For him, the entire training would have been in vain if he adopts the skeptic's arguments - even if he had acknowledged their veracity. In addition, he would have to deal with the fact that he has been giving ineffective sugar pills to patients in his practice over all these years. So, there's a lot at stake for such a homeopath. (Herrmann 2013)

Indeed there is a lot at stake. I know this as well as anyone. Almost every day, patients left my practice and convincingly explained that they had been feeling better since the beginning of their homeopathic treatment; adults, children, people with mild and severe illnesses alike. Why give all this up? Why question it? The problem for us as homeopaths is that we have not yet been able to provide an unequivocal or even comprehensible logical explanation for the claimed efficacy. We tell everyone about the patients we have healed, including our children and close and distant acquaintances and relatives, but we can't show a single study that *really* confirms these healing successes (or one that proves *with certainty* that there has been anything more than a placebo effect). Furthermore, the problem remains that our desire to speak of a healing energy in the globules is untenable if we want to use homeopathy as part of medicine, and thus also as part of natural science. We cannot simply declare the presence of this energy as a fact, but at best merely as an idea that we homeopaths have long been fond of. The same applies to the vital force. We must say goodbye to these ideas and concepts. We must also say goodbye to the globules as carriers of a substantial (material) or spirit-like effect. No further studies are required for this purpose. What is not possible in the natural sciences cannot and does not have to be proven within medicine. This part of homeopathy is simply untrue. Please read the sources for yourself.

Acknowledging this was a huge disappointment for me, and perhaps you now feel the same way.

> Intellectual honesty means that one does not claim to know (or to be able to know) something that one cannot know, and nevertheless be possessed with an unconditional respect for truth and knowledge, even when it comes to self-knowledge, and even when self-knowledge is not always accompanied by good feelings. Metzinger (2013)

This principle has been a guide to me in my investigation, and I readily admit that I was often haunted by bad feelings when I read the studies, sources, and sceptical blogs. It is always difficult to distance ourselves from cherished long-standing ideas, especially when we realize that our hopeful convictions may be replaced by a rather bleak outlook. The renunciation of vital force and its influence through energetic information in the globules is a profound step in this direction. This is particularly difficult when we become aware that as homeopaths we have nothing more to offer our patients (and that we never had anything more to offer them). But is that really the case?

Despite all the justified criticism, I still consider the amount of time and the openness with which we as homeopaths approach our patients, our differentiated care over each individual patient, and the idea of individualizing medication to be important advantages of homeopathy even today. In addition, I find the homeopathic picture of patients and disease to be much truer than those generated on a daily basis in normal medicine.

I had my patients rate me on an Internet comparison forum for physicians and I received positive feedback for the time I took, my accessibility, the conversations that seemed to have a healing effect (per se), my empathy, and so on. Not a single patient wrote to thank me for finally finding the right globules for them. Indeed, most patients do not seem to attach much importance to the homeopathic globules, but rather to the overall setting and the alternative to normal medical practice. Here we should examine which parts of this are painfully absent in science-based medicine, and which we might be able to take up and integrate from homeopathy.

Of course, science is not everything, and we homeopaths like to mention this again and again. But medicine is part of science. If homeopathy wants to be part of medicine, it must adhere to scientific principles and be evaluated according to the criteria of medicine and science. The comforting thing is that there is no further need to wait and hope for a miracle. All the facts are already on the table.

Would you entrust your life to a surgeon who says: "I don't know exactly what I'm doing here, I can't explain it exactly, and my approach contradicts all known and plausible scientific results. However, I have already been able to help some other patients, even though I cannot prove this."?

I don't want to treat my patients like that. I sincerely hope that you will be able to recognize the purpose of my book - and that it will enable us to find new ideas for a truly coherent application of homeopathy as a method that does justice to the complexity of the human being - as well as to the logic of science. As a doctor, I only want to offer my patients a concept that is consistent, correct, and checked. In the best case, I hope that my book will stimulate serious research that will enrich both homeopathy and scientific medicine. After all, there is a distinct possibility that we do not know everything yet. However, it is important to distinguish clearly between knowledge on the one hand and faith or individual personal experiences on the other, and to clearly understand the consequences of this knowledge. Even if it means saying farewell to homeopathy.

References

Berger C et al (2013) Mental imagery changes multisensory perception. Curr Biol 23 (14):1367–1372

Brien S et al (2010) Homeopathy has clinical benefits in rheumatoid arthritis patients that are attributable to the consultation process but not the homeopathic remedy:

a randomized controlled clinical trial. Rheumatology (2011) 50(6):1070–1082. https://doi.org/10.1093/rheumatology/keq234. Accessed 28 Feb 2018

Burrer E (2013) Die Kybernetik in der Medizin und Psychotherapie. (Kybernetics in Medicine and Psychotherapy). http://sigma-akademie.de/artikel/interaktive-medizin/ (in German)

Hahnemann S (1997) Organon of medicine, 6th edn (Translation Boericke). http://www.homeopathyhome.com/reference/organon/organon.html. Accessed 20 April 2018

Henzen C (2004) Glukokortikoide in Stresssituationen. (Glucocorticoids in stress situations). Swiss Med Forum 4:1187–1191 (in German)

Herrmann S (2013) Starrköpfe überzeugen. (How to convince stubborn minds). Rowohlt, Reinbek bei Hamburg. (in German)

Kirschbaum C (2001) Das Stresshormon Cortisol – ein Bindeglied zwischen Psyche und Soma? (The stress hormone cortisol - a link between psyche and Soma?) In: Lorenz R (2004) Salutogenese – Grundwissen für Psychologen, Mediziner, Gesundheits- und Pflegewissenschaftler. (Salutogenesis - basic knowledge for psychologists, physicians, health and nursing scientists). Reinhardt, München (in German)

Metzinger T (2013) Spiritualität und intellektuelle Redlichkeit. Ein Versuch. (Spirituality and intellectual honesty. An attempt). Selbstverlag, Mainz (in German). http://www.philosophie.uni-mainz.de/Dateien/Metzinger_SIR_2013.pdf. Accessed 18 April 2018

6

Here and Now

The previous chapter was actually the last in the German edition. However, for several reasons, I feel the need to include another chapter in this edition, which will focus on the here and now.

6.1 One Call - and No Response

When this book was first published in 2015, it was primarily intended as an invitation to my homeopathic colleagues at the time to take a look at my theses, which had occupied me so much and led me so far away from my own life plan. I had imagined that it should be a matter of general interest to find out to what extent homeopathy was objectively untenable - as a specific drug therapy without plausibility and without proof of effect - and what could possibly justify it further as an independent method. It was my opinion at the time that what I had found out for myself - as a convinced homeopath - and what I had set myself as a task would surely move other homeopaths in the same way. Although I had already drawn my own personal conclusions when the book was published and no longer wanted to offer my patients therapies posited on such unstable foundations, I still had the idea that an open discussion could lead to a reorientation and thus enable homeopathy to continue in some form as a method within medicine. I hoped my theses would contribute to this.

But, far from it! There was simply no such response. On the contrary, the response I actually got was a completely different one. Very soon after the publication of the book, and this is still the case today, the vast majority of

© Springer Nature Switzerland AG 2019
N. Grams, *Homeopathy Reconsidered*,
https://doi.org/10.1007/978-3-030-00509-2_6

homeopaths, whether individual therapists, professional organisations, or interest groups, have all flatly rejected my theses and suggestions, seeing them as some kind of heresy, and above all they have insisted on their position that they can offer a specifically effective drug therapy.

A constructive dialogue thus became impossible. The idea of reconsidering homeopathy within the homeopathic scene, reflecting on those of its aspects that might possibly be worth preserving in the interests of the patient, had been blocked.

6.2   The Way Forward

But this is exactly what prompted and ultimately forced me to develop my own thoughts on this matter as laid out in this book. I was fortunate enough to get into a dialogue with critics of homeopathy from the medical and scientific fields, with medical journalists, authors, and other critically minded people. In the spring of 2016, this dialogue led to the establishment of the German "Homeopathy Information Network", which I am still running today and which is committed to inform people about homeopathy, its principles, its contradictions and incompatibilities, and its classification in the medical and scientific context in general. The even deeper insights I have gained in this context, through further studies and through a multitude of dialogues and discussions have shown me to what extent I was actually still connected to homeopathy through this book. In a way, despite everything, the book was a genuine apology for homeopathy.

In this respect I have sharpened my knowledge and awareness of the fact that there is no basis for believing that any specific medicinal effect can be associated with homeopathic remedies. My current knowledge of the scientific study of the possible effects of homeopathy and of the theses that are traded as "basic homeopathic research" confirm my former complete rejection of any specific medicinal effect due to homeopathic methods. In my view, there can be no reasonable doubt about this rejection. Homeopathy as a drug therapy based on Hahnemann's ideas is unquestionably obsolete from the scientific point of view and no further research efforts are required.

6.3 The "Model" of Reconsidered Homeopathy - Quo Vadis?

And what about my thoughts for a "reconsidered" homeopathy which I was at pains to work out in the original book, namely, a method conceived out of the homeopath's detailed consultations and generally attentive attitude toward the patient?

In retrospect, it seems strange that it is precisely the sensation method with its stronger emphasis on the idea of a "core sensation", the spiritual level of illness, that has broadened my view of the aspects that I emphasize in my book as those worth preserving. But this is exactly why it deserves the place in my book, because, without the stimulation provided by the sensation method, my attention would perhaps not have been so strongly focused on dealing with aspects of the patient outside the purely physical level. Nevertheless, like so much else, the sensation method basically became an "empty shell". I had come to this method because I was looking for a "firm hold", a comprehensible method for identifying drugs. The apparent systematization of this problem in Sankaran's method, and what appeared to be a reliable "tool", blinded me to the fact that no solution to any problem is thereby provided, only a change of scene. And this applies to many variants and doctrines of homeopathy in a similar way, as I am now fully aware today. My thoughts at that time on what could be worth preserving about homeopathy are quite correct and I still see confirmation of this, but do we need any recourse to homeopathy and its methods to achieve this?

Will it be possible to assume, as I have in this book, that it is realistic to pursue a basically imaginable "conversion" of homeopathy into a conversation-centred, psychological-psychosomatic method with an occupational profile that has changed in the direction of "health management"? I don't think so today. First, homeopaths themselves show no willingness to follow this route. As a rule, homeopaths are not trained to carry out psychotherapeutic-psychosomatic therapy, not even low-threshold therapy, with the patient. Some methods, such as the Sankaran sensation method, may go in this direction, but those who, like the vast majority of homeopaths, refer to a specific medicinal efficacy as the core of homeopathy - and thus stick with the homeopathic anamnesis (for the purpose of drug discovery) rather than psychosomatic therapy - hardly offer favourable conditions for reorienting themselves towards an environment based on conversation therapy. None of the psychological/psychosocial "feedback" that currently arises in the context of homeopathy is really well-founded in a professional sense, being predominantly unsystematic and random (and sometimes just ideologically biased). This is of course far from sufficient to derive a real basis for the continued existence of homeopathy in the context of modern medicine, and is indeed questionable for several reasons.

6.4 From " Reconsidered " Homeopathy to Modern Medicine

However, as I now know, we don't need any approaches from homeopathy here. Such approaches have long been the subject of medical-scientific research; intensive work is being done on ways and means of making them usable for the practice of scientific medicine. What I did not know when writing the book was the high level of scientific research on psychosomatic medicine, on placebo research and related fields, and on the further development of the concept of evidence-based medicine.

My knowledge, assessment, and criticism of homeopathy are now much more advanced than when I wrote the German edition of my book. I still believe that the model I presented there, according to which a "reconsidered" conversation-centred homeopathy could be examined scientifically for validity, is relevant and appropriate to my considerations at that time. However, these considerations are already reflected in various fields of scientific medical research, which are already following the path I so laboriously tried to take in my departure from homeopathy in its original form.

Psychology is constantly evolving, and psychosomatics, the branch of research that researches psychosomatic connections and also evaluates medically relevant psychotropic methods in order to develop possibilities relevant to everyday practice, is in my opinion the one that will in the future take on the greatest significance in the context of scientific medicine. Then there are fields such as salutogenesis, a concept concerned with maintaining and becoming healthy, which presents itself as a "holistic" approach, just as homeopathy proposes without being able to fulfil its promise. All this is happening against the background of evidence-based medicine, which combines the scientific foundation provided by proven, evidence-based methods with medical skills and experience on the one hand and the individual needs and ideas of the patient on the other, with a view to achieving an individual treatment concept based entirely on the idea of an "emergence" - a "holistic approach" *par excellence*.

The conditions which would allow such a synthesis to come about are a completely different matter. Here, health policy can no longer keep pace with scientific findings. The demands on decision-making levels in the hierarchy will increase considerably in the future, and it will not only be a matter of material resources, but also a problem of keeping pace with medical developments in the implementation of the health care system. This issues here range from designing and continuously updating studies at medical faculties to establishing the principle of evidence-based medicine on a broad front in daily practice.

6.5 Looking Forward

And so today I stand well outside homeopathy and have become a much sharper critic of the method than I was originally when I wrote this book, and even when I finished it. It is an exclusively personal decision if someone should choose to make use of homeopathy on their own responsibility - by which I mean in an informed way, if they themselves think it is right - and I would nqt wish to interfere with that. But this should not happen at the expense of the insurance community, or under the cover of a legitimised form of therapy, as is still the case in Germany; and above all not as part of scientific medicine and the medical practices associated with it.

More than anything else, I am fighting against the non-information and disinformation of patients and consumers, against the semblance of credibility of homeopathy, which for many reasons it still enjoys on a broad front. When one hears that homeopathy is "becoming more and more popular", this is certainly not an argument in favour of it, but rather a proof that the false propaganda for "gentle and natural healing free of side-effects" is still spreading. And we know from sociological research that an opinion, if it exceeds a "critical mass" in its dissemination, will tend to spread further as a perceived truth. Personally, I have no doubt that this is diametrically opposed to any form of health care that is committed to the well-being of the patient. In this respect, I do not consider it to be a private matter whether we react against unscientific pseudo-methods, but rather a duty of public health policy.

And there is another aspect I see today with greater sharpness than when I was in the process of detaching myself from homeopathy: the potential risks of homeopathy. At that time, I trusted as a matter of course that homeopathic therapists, especially the medical doctors among them, do as a rule really know their method and its limitations, while always remaining aware of the scientific foundations of their profession and how to use them. Unfortunately, many reports have led me to seriously doubt this. The image is much darker than I had previously perceived it. But it is also clear that the therapist who "believes" in the specific effectiveness of homeopathy will not be immune to crossing boundaries, beyond which there can no longer be any talk of a responsible treatment of the patient. Such a therapist will also be subject, perhaps even more so than the patient, to the confirmation bias, i.e., the mistake of only perceiving and rating highly information that would appear to support his own view. I experienced this with myself. And unfortunately, homeopaths hardly ever recognize that individual experiences - not even those of extensive practice and several decades of practice - are worthless for the purposes of proof in the scientific sense – an epistemological triviality.

Although I have move a long way from homeopathy, I still feel committed to the same basic idea that prompted me to write this book: a commitment to better medicine, with benefits for the patient. The question of empathic medical conversation has great importance here. Today, however, I know much more about what will still be required to incorporate such a concept, and I am much more aware that improvements in the health system depend on many different factors. With this additional knowledge and the experience I have gained by travelling so far, I have written a new book: "*Gesundheit - Ein Buch nicht ohne Nebenwirkungen*" (*Health - A book not without side effects*), which was published in Germany in autumn 2017. Homeopathy now only occupies a small part of it.

I do not think this diminishes the relevance of the book you have just read, which may perhaps take you beyond pure information and show you a way out of a narrow, almost religious fixation on homeopathy to an objective view of the facts. And, of course, it should show you my core concern, namely, a medicine that is oriented toward the state of science, the therapist's skills and experience, and the needs and ideas of the individual patient: From "homeopathy reconsidered" to "medicine reconsidered".

www.netzwerk-homoeopathie.eu
www.homöopedia.eu
www.skeptiker.de
www.natalie-grams.de

Printed in the United States
By Bookmasters